Y0-CXR-630

ENVIRONMENTAL INFORMATION IN DEVELOPING NATIONS

Recent Titles in
Contributions in Librarianship and Information Science
Series Editor: Paul Wasserman

ENVIRONMENTAL INFORMATION IN DEVELOPING NATIONS

Politics and Policies

Anna da Soledade Vieira

CONTRIBUTIONS IN LIBRARIANSHIP AND INFORMATION SCIENCE, NUMBER 51

GREENWOOD PRESS
WESTPORT, CONNECTICUT · LONDON, ENGLAND

Library of Congress Cataloging in Publication Data

Vieira, Anna da Soledade.
  Environmental information in developing nations.

  (Contributions in librarianship and information
science, ISSN 0084-9243 ; no. 51)
  Bibliography: p.
  Includes index.
  1. Environmental policy—Information services—
Developing countries. 2. Environmental protection—
Information services—Developing countries. I. Title.
II. Series.
HC59.72.E5V53 1985   363.7′056′091724   84-10728
ISBN 0-313-23432-9 (lib. bdg.)

Library of Congress Catalog Card Number: 84-10728
ISBN: 0-313-23432-9
ISSN: 0084-9243

First published in 1985

Greenwood Press
A division of Congressional Information Service, Inc.
88 Post Road West
Westport, Connecticut 06881

Printed in the United States of America

10 9 8 7 6 5 4 3 2 1

To my mother and Thiago,
my links between past and future
in the respect for life

To Mary Bos,
a praise to life
(in memoriam)

Of all things in the world, people are the most precious.—
Declaration of the UN Conference on the Human Environment

Contents

Preface

This work represents six years of research and a whole life of experience in underdevelopment. More than any degree or theoretical knowledge, my own living conditions in Brazil as a member of a peasant family authorizes me to approach the theme of this book.

The research began with a focus on Brazil and the scope has been extended to a more general discussion. In developing the two phases of my studies, I have benefitted from invaluable help: the friendly dialogue with many individuals and the support of some institutions.

I wish to express my sincere appreciation to the Brazilian Coordination Program for Training of Higher Education Personnel (CAPES), the Brazilian Council of Scientific and Technological Development (CNPq), the University of Minas Gerais (Brazil) (UFMG) and the University of Paraíba (Brazil) (UFPb). CAPES at the Brazilian Ministry of Education and CNPq awarded me grants that made this study possible in both steps. UFMG and UFPb, my successive employers from 1978 to 1984, allowed me the minimum local condition to perform this research.

I would like to express special gratitude to all colleagues in the library profession and to different professionals from organizations in the fields of environment and development, in Brazil and abroad, who helped me by contributing information and motivation for this research. I am particularly indebted to those experts mentioned in the final note to chapter 3 as the most valuable sources of information for the present study. Another group of individuals—colleagues and friends—also deserve to be mentioned, since their suggestions have been essential to the accomplishment of this work: Maria Lucia Andrade Garcia, Afranio de Carvalho Aguiar, Dr. Naftale Katz, Stuart Burchel, Professor Glaucio A. Dillon Soares and Professor Peter Havard-Williams. The patient help of my friends David Spiller, Jonathan Coe, Richard Munro and Dr. Isis Paim in reviewing the manuscript was crucial to make this piece of work more intelligible. I owe Professor Marta Dosa gratitude for having initially encouraged me to embark upon research on the environment and later to have presented the manuscript to the editor.

Noteworthy was the guidance and information, through documents and interviews, granted me by the UN Environment Program (UNEP), the Brazilian Secretariat of the Environment (SEMA), the Brazilian Ministry of Foreign Affairs, the Indian Department of Environment (DOE), the Environmental Information Office of the Egyptian Academy of Scientific Research and Technology and by the Mexican Subsecretariat for the Improvement of Environment (SMA).

An important part was also played in the project by my family, my friends and my students, whose help, support, affection and encouragement I deeply appreciate.

Ecodevelopment has been adopted as the ideology guiding the present work because I understand that this concept represents an ideal balance between humankind and nature, and it is intended to diminish the inequalities among nations. In choosing such a position, I express my belief in a future where nature will be respected, nations will coexist peacefully and individuals will live as real human beings.

Acronyms and Abbreviations

ABES	Associação Brasileira de Engenharia Sanitária e Ambiental (Brazilian Association of Sanitary and Environmental Engineering)
ACAST	UN Advisory Committee on the Application of Science and Technology to Development
AGAPAN	Associação Gaúcha de Proteção ao Ambiente Natural (Gaucha Association for the Protection of Natural Environment—Brazil)
AGRINTER	Sistema Interamericano de Información para las Ciencias Agrícolas (Inter-American System of Agricultural Information)
AID	See USAID
APPLE	Ariane Passenger Payload Experiment
ASFIS	Aquatic Sciences and Fisheries Information System
AT	Appropriate Technology or Alternative Technology
BIREME	Regional Library of Medicine and Health Sciences (São Paulo)
CAPES	Coordenação de Aperfeiçoamento de Pessoal de Ensino Superior (Brazilian Coordination Program for Training of Higher Education Personnel)
CAPMAS	Central Agency for Public Mobilization and Statistics
CCMS	NATO's Committee on the Challenges of Modern Society
CEB	Comunidades Eclesiais de Base (Popular Communities Based on the Brazilian Catholic Church)
CELADE	Centro Latinoamericano de Demografia (Latin American Center of Demography)
CEMAT	Centro Mesoamericano de Estudios sobre Tecnología Apro-

	piada (Meso-American Center of Studies about Appropriate Technology)
CENAGRI	Centro Nacional de Informação e Documentação Agrícola (Brazilian Center of Agricultural Information and Documentation)
CENIDS	Centro Nacional de Información y Documentación en Salud (Mexican Center of Health Information and Documentation)
CENPES	Centro de Pesquisa e Desenvolvimento Leopoldo A.M. de Mello-Petrobrás (R & D Center Leopoldo A.M. de Mello-Petrobrás-Brazil)
CEPAL	See ECLA
CEPED	Centro de Pesquisas e Desenvolvimento (Brazilian R&D Center)
CEPIS	Centro Panamericano de Información y Documentación en Ingeniería Sanitária y Ciencias Ambientales (Pan-American Center of Information and Documentation on Sanitary Engineering and Environmental Sciences)
CERDAS	Center for the Coordination of Social Sciences Research and Documentation in Africa South of the Sahara
CETEC	Fundação Centro Tecnológico de Minas Gerais (Foundation of Technological Center of Minas Gerais—Brazil)
CETESB	Companhia de Tecnologia de Saneamento Ambiental (Company of Environmental Sanitation Technology—Brazil)
CIAT	Centro Interamericano de Agricultura Tropical (Inter-American Center of Tropical Agriculture)
CICA	Centro de Informações em Ciências Ambientais da Universidade Federal do Rio Grande do Sul (UFRS) (Environmental Information Center of the UFRS—Brazil)
CIDST	EEC's Committee of Scientific and Technical Information
CIFCA	Centro Internacional de Formación en Ciencias Ambientales (International Center for Training in Environmental Sciences)
CIN	Centro de Informações Nucleares (Nuclear Information Center—Brazil)
CIRISCA	Centro de Información y Referencia en Ingeriería Sanitaria y Control del Ambiente (Information and Referral Center in Sanitary and Environmental Engineering)
CIS	Centre International d'Information sur la Sécurité et Hygiène du Travail (International Center of Information about the Security and Hygiene of Labor)
CLADES	Centro Latinoamericano de Documentación Económica y Social (Latin American Center of Economic and Social Documentation)

CNBB	Conferência Nacional dos Bispos do Brasil (National Committee of Brazilian Bishops)
CNEN	Comissão Nacional de Energia Nuclear (Brazilian Commission of Nuclear Energy)
CNPq	Conselho Nacional do Desenvolvimento Científico e Tecnológico (Brazilian Council of Scientific and Technological Development)
CPO	Centro de Informaçao sobre Politica Científica e Tecnológica (Information Center on Scientific and Technological Policy)
CONACYT	Consejo Nacional de Ciencia y Tecnología (Mexican Council of Science and Technology)
CSE	Center for Science and Environment
CSIR	Council of Scientific and Industrial Research
DHN	Departamento de Hidrografia e Navegação (Brazilian Department of Hydrography and Navigation)
DIFRID	*Directorio de Fuentes y Recursos de Información Documental* (Directory of Sources on Bibliographic Information)
DNAEE	Departamento Nacional de Águas e Energia Elétrica (Brazilian Department of Water and Electrical Energy)
DNOS	Departamento Nacional de Obras e Saneamento (Brazilian Department of Sanitary Works)
DNPM	Departamento Nacional de Produção Mineral (Brazilian Department of Mineral Production)
DNU	Departamento das Nacões Unidas (Department concerning the UN Affairs at the Brazilian Ministry of Foreign Affairs)
DOCPAL	Sistema de Documentación sobre Populación en America Latina (Documentation System on Population in Latin America)
DOCPOP	Sistema de Documentação sobre População no Brasil (Documentation System on Population in Brazil)
DOE	Department of Environment (India)
DRTC	Documentation Research and Training Center
DST	Department of Science and Technology (India)
ECA	UN Economic Commission for Africa
ECDIN	Environmental Chemicals Data and Information Network
ECLA	UN Economic Commission for Latin America
ECO	Centro Panamericano de Ecología Humana y Salud (Pan-American Center of Human Ecology and Health)
ECOSOC	UN Economic and Social Council
ECWA	UN Economic Commission for Western Asia
EEC	European Economic Community
ELC	Environmental Liaison Center

ELIS	Environmental Law Information System
EMBRAPA	Empresa Brasileira de Pesquisa Agropecuária (Brazilian Organization of Agricultural and Cattle Breeding Research)
EMIN	Environmental Management Information Network
ENVIS	Environmental Information System
EPA	U.S. Environmental Protection Agency
EPIG	CIDST's Environmental Protection Information Group
ESCAP	UN Economic and Social Commission for Asia and the Pacific
EURONET	European Network
FAO	UN Food and Agriculture Organization
FBCN	Fundação Brasileira para a Conservação de Natureza (Brazilian Foundation for Nature Conservation)
FEEMA	Fundação Estadual de Engenharia do Meio Ambiente (Environmental Engineering Foundation of Rio de Janeiro State—Brazil)
FELAP	Federación Latinoamericana de Periodistas (Latin American Federation of Journalists)
FIBGE	Fundação Instituto Brasileiro de Geografia e Estatística (Brazilian Institute of Geography and Statistics)
FNDCT	Fundo Nacional do Desenvolvimento Científico e Tecnológico (Brazilian Fund for Scientific and Technological Development)
FWPCAA	U.S. Federal Water Pollution Control Act Amendments
GEIPOT	Empresa Brasileira de Planejamento dos Transportes (Brazilian Organization of Transportation Planning)
GEMS	Global Environmental Monitoring System
GIPME	Global Investigation of Pollution in the Maritime Environment Program
GNP	Gross National Product
GOFI	General Organization for Industrialization
HABITAT	UN Conference on Human Settlements, Vancouver, 1976
IAEA	International Atomic Energy Agency
IBDF	Instituto Brasileiro de Desenvolvimento Florestal (Brazilian Institute of Forest Development)
IBICT	Instituto Brasileiro de Informação em Ciência e Tecnologia (Brazilian Institute of Information on Science and Technology)
IBRD	International Bank for Reconstruction and Development (World Bank Group)
ICAO	International Civil Aviation Organization
ICSU	International Council of Scientific Unions
IDA	International Development Association (World Bank Group)

IDRC	International Development Research Center
IFC	International Finance Corporation (World Bank Group)
IICA	Instituto Interamericano de Cooperación para la Agricultura (Inter-American Institute of Cooperation for Agriculture)
ILO	International Labor Organization
IMCO	Intergovernmental Maritime Consultative Organization
IMEXCA	Indice Mexicano de Calidad del Aire (Mexican Index of Air Quality)
IMF	International Monetary Fund
INAMET	Instituto Nacional de Meteorologia (Brazilian Institute of Meteorology)
INDERENA	Instituto de Desarrollo de los Recursos Naturales Renovables (Institute for Renewable Natural Resources Development)
INEREB	Instituto Nacional de Investigación sobre Recursos Bióticos (National Institute of Research on Biotic Resources)
INFOPLAN	Sistema de Información para Planificación en America Latina y el Caribe (Information System for Planning in Latin America and the Caribbean)
INFOTEC	Fondo de Información y Documentación para la Industria (Center of Information and Documentation for Industry)
INFOTERRA	International Referral System for Sources of Environmental Information
INIS	International Nuclear Information System of the International Atomic Energy Agency
INPE	Instituto de Pesquisa Espacial (Brazilian Institute of Space Research)
INSAT	Indian National Satellite System for Telecommunication, Television and Meteorology
INSDOC	Indian National Scientific Documentation Center
INT	Instituto Nacional de Tecnologia (Brazilian Institute of Technology)
INTERDATA	Rede de Transmissão Internacional de Dados (Brazilian Network of International Data Transmission)
INTIB	Industrial and Technological Information Bank
IOC	Intergovernmental Oceanographic Commission
IPT	Instituto de Pesquisa Tecnológica (Brazilian Institute of Technological Research)
IRPTC	International Register of Potentially Toxic Chemicals
ITDG	Intermediate Technology Development Group
IUCN	International Union for Conservation of Nature and Natural Resources
JNU	Jawaharlal Nehru University

KSSP   Kerala Sastra Sahitya Parishad
LATINAH   Red Latinoamericana de Información sobre Asentamientos
                Humanos (Latin America Information Network on Human
                Settlements)
LCC   Laboratório de Computação Científica do CNPq (Laboratory
                of Scientific Data Processing of CNPq—Brazil)
MAB   Man and the Biosphere Program
MEDI   Marine Environment Data Information System
MIC   Ministério da Indústria e Comércio do Brasil (Brazilian
                Ministry of Industry and Commerce)
MINTER   Ministério do Interior (Brazilian Ministry of the Interior)
MIS   Management Information System
MME   Ministério das Minas e Energia (Brazilian Ministry of Mines
                and Energy)
NATO   North Atlantic Treaty Organization
NEERI   National Environmental Engineering Research Institute
NGO   Nongovernmental Organization
NIDOC   National Information and Documentation Center
NISSAT   National Information System for Science and Technology
NRDC   National Research Development Corporation of India
OAS   Organization of American States
OAU   Organization of African Unity
OECD   Organization for Economic Cooperation and Development
PADIS   Pan-African Documentation and Information System
PAHO   Pan-American Health Organization
PBDCT   Plano Básico de Desenvolvimento Científico e Tecnológico
                (Brazilian Basic Plan of Scientific and Technological Devel-
                opment)
PDS   Partido Democrático Social (Democratic and Social Party—
                Brazil)
PETROBRÁS   Petróleo Brasileiro S.A. (Brazilian Petrol Co.)
PID   Publications and Information Directorate
PLANALSUCAR Programa Nacional de Melhoramento da Cana-de-açurar
                (Brazilian National Program for the Betterment of Sugar Cane)
PMDB   Partido da Movimento Democratico Brasileiro (Brazilian
                Democratic Movement Party)
PND   Plano Nacional de Desenvolvimento (Brazilian Plan of De-
                velopment)
POPIN   Population International Network
PRODASEN   Serviço de Processamento de Dados do Senado Federal (Data
                Processing Service of Brazilian Senate)
REPIDISCA   Red Panamericana de Información y Documentación en In-
                geniería Sanitária y Ciencias del Ambiente (Pan-American

	Network for Information and Documentation on Sanitary Engineering and Environmental Sciences)
ROLA	See UNEP/ROLA
SACC	Science Advisory Committee for the Cabinet
SAHOP	Secretaria de Asentamientos Humanos y Obras Publicas (Mexican Secretariat of Human Settlements and Public Works)
SARH	Secretaria de Agricultura y Recursos Hidraulicos (Mexican Secretariat of Agriculture and Water Resources)
SCOPE	ICSU's Scientific Committee on Problems of the Environment
SEADE	Fundação Sistema Estadual de Análise de Dados (São Paulo State System of Data Analysis Foundation—Brazil)
SECOBI	Servicio de Comunicación a Bancos de Información (Communication Service with Information Banks)
SEDOC	Servicio de Documentación y Comunicación Rural (Rural Communication and Documentation Service)
SELAVIT	Latin American and Asian Grassroots Housing Service
SEMA	Secretaria Especial de Meio Ambiente (Brazilian Secretariat of the Environment)
SERPRO	Serviço Federal de Processamento de Dados (Brazilian Data Processing Service)
SICA	Sistema de Información sobre Calidad del Agua (Information System on Water Quality)
SICTEX	Sistema de Informação Científica e Tecnológica do Exterior (Brazilian Science and Technology Information System for Abroad)
SIJUR	Sistema de Informações Jurídicas do Senado Federal (Law Information System of Brazilian Senate)
SIMA	Sistema de Informações para o Meio Ambiente (SEMA's Environmental Information System—Brazil)
SINDU	Servicio Interamericano de Información sobre Desarrollo Urbano (Inter-American Information Service on Urban Development)
SINIMA	Sistema Nacional de Informação para o Meio Ambiente (Brazilian System of Environmental Information)
SIVA	Sistema de Información sobre los Usos del Agua (Information System on Water Utilization)
SMA	Subsecretaria de Mejoramiento del Ambiente (Mexican Subsecretariat for the Improvement of Environment)
SRD	Serviço de Referência Documentária (SEMA's Bibliography Information Service—Brazil)
SRL	Serviço de Referência Legislativa (SEMA's Environmental Law Information Service—Brazil)

STI	Secretaria de Tecnologia Industrial (Brazilian Secretariat of Industrial Technology)
SUPREN	Superintendência de Recursos Naturais e Meio Ambiente (Brazilian Coordination of Natural Resources and Environment)
TYMNET	Timeshare Network
UFMG	Universidade Federal de Minas Gerais (University of Minas Gerais—Brazil)
UFPb	Universidade Federal da Paraíba (University of Paraíba—Brazil)
UN	United Nations
UNAM	Universidad Nacional Autónoma de México (Mexican National University)
UnB	Universidade de Brasília (University of Brasília)
UNCTAD	UN Conference on Trade and Development
UNDP	UN Development Program
UNDRO	UN Disaster Relief Coordinator
UNEP	UN Environment Program
UNEP/ROLA	UNEP Regional Office for Latin America
UNESCO	UN Educational, Scientific and Cultural Organization
UNIDO	UN Industrial Development Organization
UNIR	Unidade Referencial - IBICT (IBICT's Referral Unit)
USAID	U.S. Agency for International Development
WHO	World Health Organization
WIPO	World Intellectual Property Organization
WMO	World Meteorological Organization
WWF	World Wildlife Fund
WWW	World Weather Watch

ENVIRONMENTAL
INFORMATION
IN
DEVELOPING
NATIONS

1.
Environmental Pollution and Development

GETTING THE CONNECTIONS

Only recently have humans shown much concern for the quality of their environment. For thousands of years they have been quite happy to use natural resources without any thought for the effect they were having upon their world.

R. Dubos[1] and J. Dorst[2] have traced the evolution of environmental disturbance from early times. According to them, man's impact on nature began with his appearance on Earth, as he had to hunt and fish for his survival. But the historical roots of environmental pollution date back 8,000 years, when men became shepherds and started using fire to make open space for grazing. The use of such a powerful implement intensified 5,000 years ago, when he adopted an agricultural economy, using fire as the main method to clear the ground and to transform forests into arable land.

Some of the practices introduced by the transformation of men into farmers are still present today, such as the use of fire, shifting cultivation, monoculture and misuse of fertilizers and pesticides. Many negative results are connected with this traditional agriculture: soil erosion and impoverishment, deterioration of the atmosphere, and climatological changes have been the consequences of burning forests and fields. Inadequate or excessive use of fertilizers has caused both the exhaustion of the soil and the eutrophication of fresh water. The massive use of pesticides has been characterized as a serious ecologic hazard with additional consequences for the future, as it interferes with the life cycle. Monoculture has invariably resulted in the exhaustion of soil and the multiplication of pests because of the favorable conditions it creates for them.

Nevertheless, it was only after the Industrial Revolution, which began in Europe at the end of the eighteenth century, that pollution became a significant problem, demanding the intervention of the state in the form of governmental agencies deriving authority from legislation. The consequent technological and social changes resulted in new roles to be played by man in society and in demands for a social structure suitable for urban nuclei, besides uncovering the appalling condition of the population's general health: pollution generated by the new technology combined with malnutrition, poor sanitation and lack of an urban infrastructure. Such a negative combination led to a savage deterioration in the environment.

In industrialized Europe, these conditions have, for the most part, been eliminated. But in most nations in the world—the so-called Third World—they persist.

With the establishment of modern forms of civil government, beginning in the industrialized nations, official programs came into being that considered the environment only vaguely as part of the public health function of administration and legislation, which provides for water supply, waste collection and disposal, housing, sanitation and vector control. Even these were considered and provided for as independent health services.

In the early 1960s, the protection of the physical environment became an important issue in government programs, in scientific literature and in the lives of people all around the world. However, it was only in the seventies that environmental control received serious and balanced consideration as a subject in its own right and was discussed in combination with two new universal problems: the energy crisis and the economic recession. Nowadays, environmental surveillance, monitoring and control are comprehensive in the developed world and include many different aspects of both the physical and the human environment, among them public health and sanitation, housing, leisure, population growth, pollution, energy, natural resources and land use. Nevertheless, the scope of government actions can vary according to the economic condition and technological development of the country and how it can cope with the complexities of balancing consumption and conservation, industrial development and a clean environment.

The discussion of information for the control of environmental degradation—hereafter referred to as pollution*—in developing countries brings together two of the most challenging problems for modern society: socioeconomic development and environmental pollution. The latter is a by-product of the two extremes of over- and underdevelopment. Any changes in technology and wealth are likely to influence the environment, insofar as the rational use of natural resources is a fundamental economic matter.

According to R. A. Easterlin, modern economic growth or development may be defined as a rapid and sustained rise in real output in the technological, economic and demographic characteristics of a society.[3] To have true development, the basic consequences of this rise should be reflected in the well-being of the individual, that is, in the improvement of his condition, employment and leisure, important facets of the human environment.

The continuous economic growth of nations, as measured by their rising gross national product (GNP) and increasing industrialization, has been cited as the cause of the deterioration of the environment by critics from the developed countries. These authors state that all nations should consider new approaches either by insisting on zero growth[4] or by using alternative technologies which could lead development back to a ''beautiful smallness''[5] or by shifting from expansion to an economic equilibrium.[6]

It is true that pollution can be a by-product of a nonappropriate technology, but it is also true that it can be a result of a wasteful pattern of consumption or bad sanitary conditions as well—all three questions of socioeconomic development.

Both socioeconomic development and environmental pollution are related to the question of the quality of life—often an elusive concept—as it relates to the level of expectation of the individual and the group, and it can vary from urban to rural life, from one culture to the next.

H. Blumenfeld puts forward a generalization that defines a good environment as one that produces healthy, happy, wise and good men

---

*''Pollution: any environmental state or manifestation which is harmful or unpleasant to life, resulting from man's failure to achieve or maintain control over the chemical, physical or biological consequences or side-effects of his scientific, industrial and social habits.'' (*Chambers dictionary of science and technology.* Edinburgh, Chambers, 1974, vol. 2, p. 914.)

and women.[7] As a common denominator it could also be said that the essence of the quality of life concept, as far as environmental planning is concerned, lies in ensuring humane living conditions. This is the point at which development meets the environment, either to improve or to lower the desired living standards.

This concept will be a point of reference in this work and will focus primarily on Brazil as a representative of developing countries. An overview of the situation in Mexico, India and Egypt, in comparison with Brazilian data, is an attempt to generalize about Third World problems and efforts for the attainment of solutions.

Throughout the study, two basic approaches will be taken. One of them will consider the consequences of industrialization and technology—namely, pollution and overexploitation of natural resources—when the protection of the environment is not a priority within an economic context. To a lesser extent it happens in developing countries, but it is principally a problem of developed nations. On the other hand, the human aspect of the problem is presented, and pollution is introduced as a by-product of poverty in developing countries and is represented mainly by poor sanitary conditions.

PHYSICAL SOURCES OF POLLUTION

Time has passed, science and technology have matured, but it seems that the basic environmental pollution problems are still the same: only their seriousness varies geographically according to the level of development of each region. Those problems can be assembled into four groups:

- Fresh water pollution, still the most prevalent problem worldwide, caused by domestic sewage, industrial disposal and agricultural wastes;
- Marine pollution resulting from ship wastes, industrial disposal and domestic sewage;
- Pollution caused by industrial effluents and the burning of fuels (cars, domestic use and agriculture);
- Land pollution by industrial, agricultural, mining and domestic disposal.

During the last three decades man has been threatened by a new type of pollution risk—radioactivity—which can affect air, water (fresh and marine) and soil and contaminate all kinds of life on earth. The contamination may be caused by atomic explosion, reactor cooling processes and nuclear waste disposal, as discussed by A. M. Weinberg[8] and P. Lebreton.[9]

A parallel aspect is sociopolitical and relates to:

- The autocratic position that governments usually take with regard to the development of nuclear technology within their own territory, denying to the public the right to information and participation in the main decisions, even where the peaceful use of nuclear energy is concerned;
- At the international level, political disputes lead governments to divert funds from socioeconomic programs to defense, in order to win the arms race.

These points make the layman suspicious of nuclear progress and motivate him to manifest his concern and annoyance about the increasing devotion of both scientists and governments to the use of radioactivity sources.

Both contamination risk and sociopolitical threats are discussed in A. D. Machado, which includes reflections from various scientists, ranging from theoretical aspects of nuclear energy generation to a discussion of the Rasmussen Report and the nuclear program of the Brazilian government.[10]

The same kind of approach is taken by G. K. McLeod[11] and E. J. Walsh[12] in their discussions of the failure of the Rasmussen Report in predicting the real risks of nuclear accidents in the United States. McLeod analyzes the question from the health point of view and the rights of society to know how radiation affects the human body and human behavior. E. J. Walsh goes further, saying that the U.S. government has lost its credibility in the handling of the Three Mile Island accident.

On the other side of the Atlantic—focusing on Windscale, England—the same suspicion of nuclear technology can be observed in the context of government-society relationships, which are discussed by B. Wynne.[13]

A more general and dramatic approach appears in J. R. Zacharias *et al.*, who discuss the danger to the world which would result from a possible nuclear war between the United States and Russia. J. R. Zacharias *et al.* propose tolerance, turning tension and conflict into creative and productive forces.[14]

There are those, however, who believe that nuclear technology carries only a potential risk and should be exploited now in order to provide benefits for mankind in, for example, engineering, power generation and medicine.

The energy crisis affecting the world since the beginning of this decade has motivated research on safe ways of exploring nuclear energy for consumption. J. F. Saglio, for instance, defends the use of radioactivity sources for this purpose and compares the environmental impact of implementation of nuclear power plants to the conventional ways of energy generation. Saglio concludes that the nature of the risks is the same, only the degree of risk varies.[15]

An extremely important aspect related to all environmental pollutants—some of these, by-products of development—is that society should have the right to choose its own destiny. In order to be in a condition to exercise this right, society must be informed of the issues and at least have health, education and freedom, basic components of the quality of life that are quite often missing in developing countries. This suggests that some conflicts in society could be at the root of environmental pollution.

NONPHYSICAL FACTORS CONTRIBUTING TO POLLUTION

Notwithstanding the importance of controlling the agents of all previously mentioned nuisances, the genesis of environmental pollution lies deeper, beyond the physical sources, and can be attributed to political, technological and social factors within a specific country.

Political Factors

The main determinant of environmental affairs is political. It reflects the ideology of the government of a specific country and the values (moral, religious and philosophical) of a particular culture.

To start with, the duality of capitalism-socialism may have a differential content as related to environment. While in a capitalist society, based on market economy, the emphasis is on mass production and consumption, private property and the pursuit of individual profit, in a socialist economy, based on central planning, the aim is to build up a common wealth, which would be shared equally by all citizens.

Depending on which of these ideologies is adopted by a government, the consequences upon the environment may vary. One consequence may be the overexploitation of both humans and nature in a capitalist country, since they are seen—from an immediatist perspective—as resources to be used in order to produce money for the highest layer in the society. On the other hand, one may find a tendency to conserve natural resources in a socialist country as part of the concept of common property owned and centrally administered by the state with public participation in work and its resulting benefits. This kind of orientation may lead citizens to care for the environment in order to share its benefits and may induce government, as the legal possessor and only administrator of the state, to be more responsible.

The concern for nature as a feature of socialism has been expressed by F. Engels in his *Introduction to Dialectics of Nature* and *Part Played by Labour in Transition from Ape to Man*, where he alerts mankind to the dangers of interfering with the traditional course of nature in order to increase production in the short term.[16]

With regard to the management of natural resources, T. dos Santos sees the idea of rationality as contingent to the social system and sees waste as an integral part of the capitalist world.[17] The same conception about capitalism is shared by C. Furtado when he identifies the obsolescence of goods as the essence of capitalism and, as an inevitable consequence, a high pressure put on natural resources.[18],[19]

A critical analysis of capitalism and socialism as they affect the environment has been written by H. Stretton. He states that the Left loses prestige in the eyes of the people because most governmental actions, both in safeguarding natural resources and in cleaning the polluted environment, are detrimental to the lower class because of the associated cost increases and scarcities of some natural commodities. He then suggests that the conservationist movement is right-wing and basically benefits the upper class.[20]

It can be argued that the interrelation between sociopolitical ideology and environmental management exists only in theory and that in the practical plane the results are not as well defined nor are they predetermined solely by the ideology. Results would also depend on the administrative and political structure of the nation, the strategic goals at that given moment, and on the availability of local information within the central government. Any of those conditions could lead either to a capitalist government protecting the environment or to the socialist government despoiling it. Thus, one may conclude that environmentalism—a belief that incorporates the concern for both humans and nature—is itself a political ideology, neither right- nor left-wing. It could be closer to the ruling principles of the City of Being proposed by E. Fromm: shifting from control over nature to control over destructive social forces and moving from authoritative business or state initiative to participation of free, responsible and well-informed citizens.[21]

Going beyond the politico-economic ideology professed by any government, two other determinants also could be considered among the political factors that provide the framework for the value ascribed to nature by man: the religious and philosophical doctrines characteristic of a specific culture. J. Dorst[2] and L. White, Jr.,[22] point out the contrast between Eastern and Western thought in what is of concern to man-nature relationships. The same ideas are shared by J. Le Goff, who shows the contrast between Christian Europe and Moslem Eastern cultures.[23]

Christianity preaches that God created everything and then created man in His own image. He allowed man to name all the other creatures, thus revealing His desire to settle man at the top of creation, as the master and consumer of nature, which has been created only for mankind's pleasure and benefit, as stated in *Genesis*: "Have dominion over the fish of the sea, and over the birds of the air, and over every living thing, that moves upon the earth."[24]

Animism was a common belief in antiquity and it is still common today among primitive tribes. However, history has shown for centuries that whenever Christian missionaries meet native groups, with their own pagan religions centered in nature, they struggle to shift this belief to the traditional anthropocentric Christian doctrine. It is possible that, once man has abandoned this special feeling for nature, he feels himself free to deploy the environment if he feels like it because he sees it merely as something destitute of spirit.

Christianity had God, man and nature reconciled only briefly, from 1150 to 1300, under the medieval scholastic doctrine. Saint Francis may be mentioned as a representative of this period. He tried to eliminate the dualism of man and nature, to restore the concept of soul to nonhuman living beings and to reestablish equality among all creatures. In the Franciscan doctrine, man and the animals are brothers, rather than master and servants.[22] More recently, Pope John Paul II included the environment as a topic in his first encyclical letter, *Redemptor Hominis*, warning man not to destroy nature through pollution and military weapons.[25]

Saint Francis's ideas share a common basis with the oriental doctrines of Buddhism, Hinduism and Taoism, which consider nature and
all living beings as proceeding from God, whose spirit they share.

In Taoism, Te gives each element of the cosmos its special character, the same way Tao—the principle of life—unifies diversity and gives
balance to the universe. Once sensitivity is a common characteristic of
all living beings, all creatures deserve the same respect. Thus, the best
form of environmental management from a Taoist viewpoint is non-
intervention (*wu-wei*) in the natural course of the life cycle.[26]

Hinduism, through its sacred books of Veda, presents the natural
world made up of interdependent elements; among these, humans and
nature are part of the same indivisible organism. Within such a variety, Rita is the principle of harmony and beauty.[27]

Eastern religions and philosophies are a unified body of doctrines,
which lead people to think, contemplate and achieve integration with
the spirit of nature. On the other hand, Western religions remain apart
from the schools of philosophy, are based on the concept of morality
and tend to encourage people to detach themselves from the physical
world, as an ideal of Christian perfection.

However, political ideology, philosophy and religion should be seen
only as indicators of tendencies toward a certain pattern of behavior of
both government and people within a certain culture in relation to the
environment. Although these considerations cannot be taken dogmati-
cally, environmental concern is more likely to exist either under a policy that leads both government and people to share the responsibility
for the environment or within a culture whose values make man and
nature allies in the issue of high quality of life.

Technological Factors

Science and technology, as conceived today, are a Western achieve-
ment. According to L. White, Jr., science began as a development of
natural theology when, in the thirteenth century, man started observing
and analyzing nature in an effort to understand God's mind by discov-
ering how His creation operates.[22]

From the eighteenth century on, scientific knowledge became progressively more independent from theology. On the other hand, technology preceded science as an empirical method and still represents
man's dominance over nature. Except for the chemical industry, which
used scientific principles in the eighteenth century, it was only by the

end of the last century that science and technology became linked. From World War I on, this link became stronger and more permanent, for better or worse, depending on one's point of view, so that one discovery could be used either for the benefit of mankind or for its perpetual harassment and final extinction.

As far as the environment is concerned, technology has been blamed both for inadequacy and for generating overdependency on natural resources. Under technological inadequacy, one can include the use of old-fashioned equipment and processes, the adoption of unsuitable processes to a specific environment and the "superefficiency" of science and technology.

The use of old-fashioned processes and equipment is a frequent threat to the environment in Third World countries. It happens either because they do not have the knowledge and the means enough to develop their own technology or because they are politically and economically subjugated by the developed countries. The sale of old-fashioned technology by rich countries to poor countries, leading to pollution, is an everyday event. It is carried out under the label of "aid" and with the excuse that the environment of the developing countries has a greater capacity to absorb pollution. An example of such a transfer of technology is proposed by the Commission of the European Communities in its program for 1977–81.[28]

For I. Illich, any technology transferred from developed nations to the Third World is inadequate and "could produce irreversible despair, because the plows of the rich can do as much harm as their swords,"[29] since they violate the physical environment, affect social structure and change cultural values.

In redefining technological risk, B. Wynne sees modern technology unviability (meaning internal incoherence and external conflict) as resulting from the alienation of society by the authoritarianism of decision makers and experts.[30] Technology has also been described as unsuitable for both man and his environment as it generates an inhuman atmosphere and violates nature. E. Schumacher says:

Technology recognises no self-limiting principles in terms, for instance, of size, speed, or violence. It therefore does not possess the virtues of being self-balancing, self-adjusting, and self-cleansing. In the subtle system of nature, technology, and in particular the super-technology of the modern world, acts like a foreign body, and there are now numerous signs of rejection.[31]

He then presents a proposal for a technology with a human face, softer to both man and nature. His basic concern is related to developing countries that have been using imported technology unsuitable to their own social and physical environment.

Resisting to this proposal, G. Bolaffi understands that it is a naive ideology that ignores the elementary requirements of modern transnational and monopolist capitalism.[32]

Considering all the political forces involved, H. Rattner criticizes people who believe alternative technology (AT) to be the key to development. He accepts AT as a fair movement toward the mobilization of the masses intended for social organization and self-reliance. So, AT would not be a solution per se but the expression of an alternative project of society.[33] Developing countries are very suspicious of this movement for intermediate technologies, fearing it is a new device to keep them backward. The answer presented by them to E. Schumacher is "small is beautiful but large is powerful."

More recently, however, the concept of alternative or appropriate technology as the one adequate for local physical and social conditions (not just an intermediate or second-class technology) has been accepted by the Third World, at least to parallel modern imported technology as one more option. An indication of this is both the inclusion of a large number of AT institutions from seventy developing countries in the UN Environment Program's (UNEP) directory[34] and the disclosure of the Organization for Economic Cooperation and Development's (OECD) research data showing that R&D expenditures on AT by developing countries are almost as high as in industrialized countries.[35] The problem facing developing countries, then, is how to become powerful and keep beautiful.

Citing the superefficiency of technology as a cause of pollution is to a certain extent a bizarre position, but it has been actually proclaimed as such. Its negative effects on the environment have been criticized by individuals and groups, for instance, the defenders of alternative technologies and some conservationists.

According to the critics, these effects are as follows.

- Technology sustains economic growth, which is in itself undesirable because it creates conditions likely to cause stress in the environment through production and consumption and because there are obvious limits to continuous growth in a finite world.[36]

- Science and technology favor unwanted population increases by improving health and sanitary conditions.
- Side effects, especially social consequences, are inseparable from new technology, many of them being unpredicted and sometimes even unpredictable.[37]
- Technology destroys human values and creates weapons for men to kill each other and ruin the environment.
- Technology is a single-minded phenomenon whose aim is development at any price.

All these concerns suggest that the postwar society has shifted from the technological myth to a mistrust and a fear of science and technology.

It is now a common belief that governments all over the world have been using science and technology as a political power to dominate one another, either peacefully or through wars. In this context, C. Furtado underlines—with fear for the future—the military orientation which now characterizes technological progress: modern inventions are increasingly being determined and funded by military budgets.[38] Within the country that has taken this technocratic approach, the situation could tend to an unbalanced distribution of resources between the technological and the social needs of the nation.

An example of such deviation is the development of a sophisticated war industry by the Brazilian government, which, in 1982, invested 5 to 10 percent of the total value of its FNDCT (the national fund for R&D in science and technology) in military research and is responsible for 45 percent of the arms exported by the Third World, which is over 15 percent of the total of Brazilian exports of manufactured goods.[39]

The above data on Brazil support UNEP's denunciation that war seems to be more important for governments than development. UNEP's objective information is that global military expenditures are twenty times as great as the "aid" given to the Third World by developed countries, and that developing countries spend four times as much on military apparatus as they receive in official development "aid."[40]

On the international scene, such an approach has led wealthy countries to the development of new powerful weapons—industrial and genetic—likely to exterminate life. These wealthy countries have also run some "development programmes," which include social, medical and technological assistance, like the Point Four created by the United States.

But these programs have been attacked for using both the population and the environment of poor countries as guinea pigs to experiment with new drugs and processes, besides exploiting their natural resources. Military "aid" is another feature of the politico-economic domination of backward countries, involving negative ecological consequences, waste of resources and loss of freedom for citizens.

To point out only one actor of this scenario, and not to be biased by citing a source from developing countries, it is worth mentioning N. Chomsky and E. S. Herman's discussion of the United States's undesirable intervention in the Third World, imposing oppressive regimes and suppressing human rights.[41] Granada, Nicaragua, Lebanon, Afghanistan, Chad and Mozambique also could be mentioned just to show that the situation unfortunately has not changed and that the actors are various, even though they all have in common the same ambition, authoritarianism and disregard for life.

A defense of technology from the point of view of a developed country has been made by M. Kranzberg, who points out technology's positive influence in all areas of human activity.[42] The reconciliation of technology and humanism in the context of Latin American development is proposed by P. Freire,[43] who, on another occasion, decried the "massification" of society by alienating technology and consumerism, suggesting that the real question should be how to avoid the mythic deviation of technology[44] and not merely stop technological development.

These arguments show that all facets of development are at the root of the environmental problem, linking together the political, technological and social factors of its disturbance.

This conclusion is related to C. Quigley's idea of conceiving the environmental crisis as a sociopolitical problem rather than as a technological one. According to him, technology itself cannot be blamed for the environmental crisis; the ecologically disruptive directions it sometimes takes is rather a symptom of society's need to reform its pattern of beliefs, values and assumptions, than a cause of disorganization.[45]

Social Factors

The core of this matter is deeply related to the distribution of wealth and world growth, the latter being considered in terms of both population and GNP increase.

The unfair distribution of wealth can be observed among nations and, within a given country, among social classes. At the international level— as shown by A. H. Westing—this is transparent if one considers that, from the global population of about 4,500 million, 22 percent of the population holds 43 percent of world land and have a per capita GNP of at least twice the global average for this value. On the other extreme, 63 percent of the world population lives in 38 percent of the global land (partially desert) and their per capita GNP is only one-half (or less) of the global average for this value. The disparity of that distribution could otherwise be pinpointed by the classification of the UN International Monetary Fund (IMF): twenty-two nations are considered developed and over ninety are not, and, among the latter group, forty-two countries are seen as extremely poor.[46] To complement the contrasting picture, two more data could be added: from those underdeveloped countries, 800 million people live in absolute poverty;[47] life expectancy in developed countries is over seventy years, but in some African and Asian countries it is still under fifty years.[48]

The world population explosion has been a constant theme in environmental literature and a concern of the international organizations dealing with social issues. Some publications have a very pessimistic tone, as in *Population Bomb*, by P. Ehrlich,[49] *Too Many*, by G. Borgstrom,[50] Malthus's classical *Essay on the Principle of Population*,[51] *Famine—1975!*, by W. Paddock and P. Paddock[52] and the more recent articles by A. Ehrlich and P. Ehrlich[53] and by A. H. Westing.[54] They all express their worries about the available resources in relation to the growing population. Almost in the same line are L. R. Brown and P. Shaw, who consider that all attempts to align society on a sustainable path will result in nothing if population growth cannot be brought to a halt.[55]

Some other authors take a more optimistic view and believe in science, technology and economic development as the solution for the world's problems. It is worth mentioning R. Taagepera's model, which contrasts with the bulk of the literature by relating population growth with technological advancement.[56] Also on population, the proceedings of the eighth annual symposium of the Eugenics Society present a reasoned and balanced discussion of the subject.[57]

The general concern about population as related to environment is that it will increase the consumption and exploitation of natural resources, will increase the production of waste and will result in greater demand upon a limited physical space. On top of this, it has been observed that the rate of population increase is higher in the developing countries for a number of possible reasons: either they do not have any family planning programs or the national health service is inefficient or the individuals themselves are not able to plan their family size because of a lack of both knowledge and resources. It is also true that in many agricultural societies parents still regard their children as a labor force and as a blessing. However, a modern view of the problem presents demographic growth as an inverse function of the socioeconomic level of the population.

Within this scenario, developing countries have first ignored population growth as a problem; then some of those countries have tried to halt it by force; and now it seems that the approach has shifted from the previous authoritarian position to getting demographic control through social development, that is, health, nutrition, housing, jobs, education and mainly by women's participation in society. This new approach is being tried through general government plans of development (including family planning as a part of health, education and welfare programs) and through the various activities of nongovernmental organizations (NGOs) and AT organizations.

Having in mind the demographic situation of developing countries, T. Vittachi, from the Public Information Division at the UN Fund for Population Activities, sees population growth, within a broad socioeconomic context, as a multiplier and intensifier of problems rather than the cause of them and recommends development as the key to population problems.[58]

Yet in the case of developing countries, the population explosion—which might not be a problem in itself—combines with the lack of technology for proper use of the environment and results in food shortages, insalubrity, malnutrition and deficiency and infectious diseases. In a cyclical natural process, it ends up by despoiling and contaminating the environment.

A parallel issue characteristic of this century is the movement of the population to large urban concentrations, as discussed by G. Bell and J. Tyrwhitt[59] and by D. Popenoe[60] with regard to its implications for both urban and rural areas, which affects the physical and social environment. The urban population of the world has been increasing at an average of 2.9 percent since 1960. In some areas this trend is even more impressive, as for instance, in Latin America, where urban population in 1970 represented 57.4 percent of the total population. In 1980 it rose to 64.7 percent. It was also found that in 1950 greater Buenos Aires was the only city in the Third World that had a population over 4 million. In 1980 there were twenty-two cities of such size in the Third World, compared with sixteen in the industrialized world.[61]

Urban concentration in developing countries becomes dramatic if one considers that a third of its population lives in squatter settlements, divorced from social benefits and the political process, despite their potential. It should be added that the slum population in the Third World increases at a rate of 10 to 15 percent yearly. In developed countries, cities represent power and affluence through technology. For the poorer people in developing countries, cities—seen from the angle of their industrialized structure—symbolize survival and social mobility through better health, job opportunities and education, which, unfortunately, most city dwellers never get to experience.

This accelerated movement to towns and cities overloads the capability of the existing local government infrastructure to provide public services for the community. As a consequence, many deformations arise in the whole system. Some of them result in deterioration of the environment, such as insufficient water and waste treatment, improper housing conditions and heavy traffic with increasing production of fumes and noise. The situation can get even worse where there happens to be a concentration of both population and industries in the same area.

This problem has been approached from two directions: by government through land-zoning policy in the new urban plans and by wealthy inhabitants moving to suburbia. Aspects of both urban and suburban human environment are discussed by H. J. Schmandt and W. Bloomberg, Jr.,[62] A. Campbell et al.[63] and W. Michelson,[64] who analyze American cases as a starting point. Socially speaking, neither of these two approaches has resolved the situation since they create an unfair division of the environment, reinforcing inequalities in the distribution of wealth: a large number of poor people share a decaying environment while a few rich people get a big share of the best nature can offer, as discussed by D. Popenoe.[60], [65]

In the context of developing countries, S. El-Shakhs discusses urban growth and recommends some broad and conventional management options, such as reduction of human fertility, migration control, regional development and land use planning.[66]

The urbanization of Latin America, as related to its style of development, is focused on by L. Pereira[67] from the sociopolitical point of view. V. Faria analyzes the Brazilian process of urbanization from 1950 to 1980, and curiously demonstrates that, in spite of the increase in absolute numbers, the rate of growth of big cities in Brazil has been decreasing in the last decades.[68] G. Bolaffi, in the same line, comes to the point of worrying about government measures to freeze the city of São Paulo's industrial growth, which is showing an already existing tendency toward stagnation. He sees this move backwards as partly caused by those controlling measures and also as a consequence of bad urban planning, which results in daily travel for workers.[69]

Different solutions have been tried for both physical and social environment problems in urban areas, but this is still an open question within the trends of modernization and development.

Unequal distribution of income and wealth is another social factor of environmental pollution. From the point of view of individuals, low income also leads people to degrade the environment unintentionally, either by dwelling in slums where they are surrounded by their own waste or by causing soil erosion through deforestation and intensive use of small plots of arable land or by contaminating the soil and water where they are not served by sewerage systems. Speaking from the point of view of countries, poverty and technological underdevelopment have the similar effects of leading to uneconomic use of natural resources, domination of backward countries by developed nations and all sorts of inequalities and discomforts for human beings.

An understanding of the negative consequences of underdevelopment motivates nations to compete for higher economic growth rates—one of the features of development—which are likely to give international political power to the country and to generate social well-being for citizens.

Economic growth measured by increments in the GNP was very fashionable during the fifties and early sixties. It corresponded also to the period when nations were running after technological advancement blindly. From the late sixties on, people's values have been changing in developed countries. Thus, as J. Robertson underscores, they started questioning the GNP as a fair measure of development since it takes into account only the exchange value of goods and services produced in the money economy but it does not consider the quality of life and human values.[70]

The desirability and feasibility of a continuous economic growth in a world with limited space and resources also have been mistrusted. Growth then began to be considered a danger to the environment because it forces the exploitation of natural resources in order to support mass production and generates more waste as a consequence of consumption and obsolescence.

This change reflected the new thought predominant in postindustrial societies and has been associated mainly with the entrance into the space age. Landing on the moon in 1969 gave man an increased consciousness that there is just one Earth, and it has limited resources. The expression ''spaceship Earth'' was in fashion, and plans to save it or to survive in it have been designed, as in P. Ehrlich and R. L. Harriman's *How to Be a Survivor*.[71]

The United States launched in 1970 the idea of ''zero growth,'' and some theorists like P. Ehrlich discussed the level of semidevelopment that poor nations could still aim for and how to de-develop the over-developed nations.[72]

A pessimistic forecast has been made by D. Meadows *et al*.[4] in their report to the Club of Rome, which anticipated famine and ecological disasters in the next hundred years as a consequence of a continuous growth in wealth, technology, population and pollution. After this publication, many other global models and forecasts were put forward—including a second report to the Club of Rome[73]—as analyzed by G. Poquet,[74] W. E. Hecox[75] and A. K. Biswas.[76] The most recent global studies that have come to the public are *Interfutures* sponsored by the OECD and *Global 2000*, prepared by the U.S. government, none of them optimistic.

The *Global 2000* study interrelates sectoral models to show that world problems are somehow universal and nations are interdependent to plan the future of resources availability, population and environmental quality. Energy, food, minerals (including water) and the environment will constitute serious problems for a poorer and highly populated world, mainly as far as less developed countries are concerned.[77]

While the above studies focus on the stress of population growth over natural resources, K. Finsterbusch's article analyzes the adverse sociopolitical effects of increasing scarcity. Devoted primarily to affluent societies, his model can be applied to all countries, suggesting that social inequalities and political authoritarianism would come as a consequence of increasing paucity. On the other hand, the study indicates that egalitarian societies have a better chance to cope with scarcity.[78]

Such a position is very important and highlights another perspective to approach global problems in relation to the availability of natural resources: the key may be in social justice and politico-economical equality, both frequently disregarded in internal policies of developing countries, and in the dialogue among nations, especially in the North-South relationship.

Amidst so many pessimistic forecasts, it is worth mentioning the study prepared by the Fundación Bariloche—*The Latin American World Model*—as the Third World's answer to the Club of Rome. It does not foresee serious limits to world growth and understands that threats to the environment would stop if we could build an egalitarian society at both national and international levels.[79]

An egalitarian international society is also the proposition made by the Brandt Report, which looks for international peace via a more just North-South relationship, throughout its political, economic and cultural aspects. This beautiful piece of political literature, prepared for the United Nations, underscores—bitterly but realistically—that the major burden for combating poverty in developing countries[80] lies with ourselves.

People in the Third World are very conscious of their burden, as it has been a constant in their history, but they demand and still hope to achieve a fairer politico-economic relationship, as spotlighted in Cancun, Mexico, by the Brazilian Minister of Foreign Affairs, R. Saraiva Guerreiro, among others. He defended—as in the Brandt Report—that world problems are interdependent; so, all nations (rich and poor) should look for common solutions as a fundamental condition for the survival of the world economy.[81]

Interdependence of the world economy and environmental problems is also the position of J. A. Lee, who recommends economic growth as the solution for global malaises, provided that such development be centered on people, be guided by environmental concern and get enough support to sustain itself.[82]

At this point the *World Conservation Strategy* should be mentioned, an impressive document prepared by the International Union for Conservation of Nature and Natural Resources (IUCN), since its central line is conservation for sustainable development. Motivated by the same worries about scarce resources for such an increasing population, it stresses the mutual dependency of conservation and development, seeing these as complementary. Hence, it identifies common lines of action for all countries and a special set of measures for the Third World to attain both development and natural resources conservation.[83]

S. D. Talisayon comes to an optimistic conclusion about the future of mankind in his model, which foresees nations adopting new ethical styles of development as a natural response to the global environmental crisis.[84] This is a quite acceptable conclusion if one observes how many people from higher classes (even in developing countries) have already adopted a new style of life, based on simplicity, self-sufficiency and healthier eating habits. Surely it will end up by softening social inequalities and alleviating environmental burdens as a kind of world self-regulating mechanism. The world's salvation through the adoption of environmental ethics and a new, ecologically sound style of life (that is, AT) is also L. R. Brown's proposition.[85]

A sound defense of growth within the environmental context, and even of the validity of GNP as a fair index, is made by W. Beckerman in *Two Cheers for the Affluent Society*[86] and in *In Defence of Economic Growth*.[87]

A new index has been sought to represent both development and quality of life. N. H. Jacoby mentions the construction of indices of change in social well-being, by the Russell Sage Foundation and by the OECD, besides the construction of an index of net national welfare by the Japanese government. All three indices try to balance people's needs with both quality of the physical environment and national economic growth.[88]

Irrespective of their sources and factors, environmental pollution problems need to be analyzed in a wide perspective because this subject presents many facets that interrelate with other aspects of the human environment. A typical example of this interrelation is brought forward by the use of pesticides. Their intensive utilization in agriculture has been discussed by environmentalists in connection with their harmful effects on human and animal life.

R. Carson's *Silent Spring* is a classic in ecological literature, which discusses the unrestricted proliferation of chemical pesticides and their use by people who ignore their potential for harm.[89] More recently, M. Linerr shows that even specialists can make the same mistake, and then blames the UN Food and Agriculture Organization's (FAO) tsetse eradication program in Africa for using toxic pesticides banned in the West and replacing wildlife by monoculture.[90]

On the other hand, the beneficial effects of pesticides in agriculture have been saving millions from hunger; and public health medicine extols their properties to exterminate the carriers of diseases, such as malaria, typhoid and encephalitis. Based on these facts, Lord Rothschild accuses R. Carson's book of being a one-sided indictment of DDT. After considering the return of malaria in certain countries because of the abolishment of that pesticide, he concludes: "Such are the risks of a no-risk society. Are we getting too clever by half? Of course not; but we must not allow the results of our cleverness to make us panic, and we must remember that one man's poison may be another man's life."[91]

What can never be forgotten when dealing with the environment is that people's life and well-being depend on it as much as its quality depends on people's care, and that the physical environment is only one link in a complex chain. It is not sensible to go from a previous conception of people on top of the universe with nature to serve them to a new one where nature becomes an untouchable deity and humans a malefic appendix.

THE IDEOLOGIES BEHIND THE SCENE

As the ecological movement became popular, people introjected the ideas produced by leading individuals from the developed world. First, those ideas are disseminated—through conferences, books and periodical articles—to national elites. Then, they are projected, as a common ideology, over mass population through various foreign and national channels, such as the media, schools and the church. This picture evidences that the implementation of such an ideology is basically a question of efficiency in communicating environmental information.

According to L. Althusser, ideology represents the imaginary relationship individuals may keep with their real living condition.[92] In the light of this concept, the above situation is a typically ideologic frame: people's perception of their real environment takes the shape of the interpretation of individuals alien to those people's reality, thanks to a powerful diffusion of information at both internal and external levels. Ideology lies in the essence of the political relationship among unequal groups—within a given country and between nations—to disguise domination and exploitation through the projection of an illusionary public interest about basic matters concerning social and physical environments.

Three basic lines can be identified in the environmental discussion throughout the literature: conservationist, developmentalist or technocratic and ecodevelopmentalist, each one corresponding to an ecological ideology. Although scientifically coherent and socially defensible, those ideologies sometimes have been blamed for masking specific interests of dominant groups or countries.

In the Third World, countries like Tanzania and Kenya have adopted conservationism as the ideology guiding governmental policy and politics, while Brazil, Mexico, Venezuela, Nigeria and Egypt could be mentioned as followers of a developmental ideology. China is a model country which has adopted ecodevelopmental ideology, being followed by a still-reluctant India.

Conservationism

This ideology is professed by the largest group of the ecological movement. Traditionally, conservationism has been characterized by a narrow line of thought, as it sees the protection of the environment as an end and blames external, isolated factors, such as technology, population increase or economic growth, for the deterioration of the environment. Using these arguments, it has created the so-called doomsday literature, which emphasizes the risks of world growth and advancement in a limited environment. There are famous scientists affiliated with this group, who are responsible for some classics, such as *Silent Spring* by R. Carson,[89] *Science and Survival*[93] and the *Closing Circle*[94] by B. Commoner, *Man Adapting*[1] and *So Human an Animal*[95] by R. Dubos, *Population Bomb* by P. Ehrlich,[49] *Population, Resources, Environment* by P. Ehrlich and A. Ehrlich,[72] *Global Ecology* by P. Ehrlich and J. P. Holdren,[96] *How to Be a Survivor* by P. Ehrlich and R. L. Harriman,[71] *Too Many* by G. Borgstrom,[50] *Ecotactics* by R. Nader[97] and *Hungry Nations* and *Famine—1975!* by W. Paddock and P. Paddock.[52]

As regards the institutions associated with this group, it is worth mentioning the Club of Rome (Italy), Friends of the Earth (England), Sierra Club (United States) and IUCN (Switzerland), all involved in publishing, research support and alerting the public. The Club of Rome's Project on the Predicament of Mankind has been made famous through the report *The Limits to Growth*,[4] prepared by a group from the Massachusetts Institute of Technology.

Developmentalism

The group of developmentalists or technocrats is formed largely of government planners, many economists and some scientists. They are the most outspoken defenders of the Third World's development and the self-determination of individual countries. In reacting against conservationists, the tone of their writings has taken another equally extreme position. While conservationists define the quality of life on the basis of a clean environment, technocrats understand it on the sole basis of development, without much concern for nature. The motto of technocrats is "produce-pollute-clean."

H. Stretton[98] and J. De Castro[99] speak as social scientists. H. Kahn,[100] E. Gerelli and J. P. Barde[101] and W. Beckerman[86], [87] represent the point of view of the economists. M. Ozório de Almeida[102] expresses some ideas shared by governments of the Third World. Nevertheless, the bulk of the technocrats' literature is represented by government documents and speeches made by representatives of developing countries at international fora, like the General Assembly of the United Nations.

Ecodevelopmentalism

This is the most recent line of the environmental movement, launched in June 1973 by Maurice Strong, when Executive Director of the United Nations Environmental Program (UNEP), during the first meeting of its Governing Council. It can be considered a product of the 1972 UN Conference on the Human Environment, at which the problems of development and environment were confronted. The concept of ecodevelopment represents a compromise between the two extreme approaches previously described, as it integrates the ecological prudence and looks forward to a development that harmonizes cultural, economic and ecological factors.[103] Thus, it has also incorporated the literature on AT.

I. Sachs has been disseminating this idea of an environmentally compatible development, and his article "Environment and Styles of Development"[104] is a comprehensive analysis of the question. In "Meio-Ambiente e Desenvolvimento: Estratégias de Harmonização,"[105] I. Sachs suggests the ways to turn ecodevelopment into a feasible practice. Interviewed in Brazil in 1979 he reviewed some achievements of the movement and compared its philosophy with that of the intermediate technology group.[106]

A theoretical approach to ecodevelopment is discussed by R. G. Wilkinson in *Poverty and Progress*, an ecological model of economic development,[107] while N. Myers analyzes China's policy as a typical example of environmentally sound development.[108] B. Ward's *Progress for a Small Planet* also can be included in this category, although she is biased toward the Western world.[109]

Even though they do not explicitly refer to ecodevelopment, two authors deserve mention as defenders of a balanced position, similar to the ecodevelopmentalists: J. Maddox, with *The Doomsday Syndrome*,[37] and A. Sauvy, with *Zero Growth?*[110] They both discuss all extremist arguments in an optimistic way and present some possible solutions for both developed and developing worlds.

J. Dorst, in *Before Nature Dies*, calls for the reconciliation of conservationists and technocrats, as regards a rational management of Earth's resources:

The ''protectors of nature'' must learn that the survival of man requires intensive agriculture and a complete transformation of certain areas; they must abandon a number of sentimental prejudices, some of which have done serious harm to the cause they are defending. On the other hand, the technocrats must admit that man cannot free himself from certain biological laws, and that a rational exploitation of natural resources does not involve transforming habitats automatically and completely. They must understand that the preservation of natural areas constitutes land-use quite as much as their modification. A realistic agreement between economists and biologists can and must lead to reasonable solutions and assure the rational development of humanity in a setting in harmony with natural laws.[111]

NOTES

1. DUBOS, R. *Man adapting*. New Haven, Conn., Yale University Press, 1967.

2. DORST, J. *Before nature dies*. London, Collins, 1971.

3. EASTERLIN, R. A. Overview. In: SILLS, D. L. (ed.). *International encyclopedia of the social sciences*. New York, Macmillan and Free Press, 1968, p. 395.

4. MEADOWS, D. H. *et al. The limits to growth: A report for the Club of Rome's project on the predicament of mankind*. New York, Universe Books, 1972.

5. SCHUMACHER, E. T. *Small is beautiful: A study of economics as if people mattered*. London, Abacus, 1973, pp. 122–33.

6. ROBERTSON, J. Towards post-industrial liberation and reconstruction. *New Universities Quarterly*, 32 (1): 6–24, Winter 1977–78.

7. BLUMENFELD, H. Criteria for judging the quality of the urban environment. In: SCHMANDT, H. J., and BLOOMBERG, W., JR. *The quality of urban life*. Beverly Hills, Calif., Sage, 1969, vol. 3, p. 139.

8. WEINBERG, A. M. Is nuclear power acceptable? *Science and Public Policy*, pp. 455–66, Oct. 1976.

9. LEBRETON, P. Les aspects écologiques des plantes nucléaires dans l'environnement côtier, *Bulletin d' Écologie*, 7(1): 33–59, 1976.

10. MACHADO, A. D. (ed.). *Energia nuclear e sociedade*. Rio de Janeiro, Paz e Terra, 1980.

11. MCLEOD, G. K. Some public health lessons from Three Mile Island: A case study in chaos. *Ambio*, 10(1): 18–23, 1981.

12. WALSH, E. J. Three Mile Island: Meltdown of democracy? *Bulletin of the Atomic Scientists*, 39(3): 57–60, Mar. 1983.

13. WYNNE, B. *Rationality and ritual: the Windscale inquiry and nuclear decisions in Britain*. Chalfont St. Giles, Bucks, British Society for the History of Science, 1982.

14. ZACHARIAS, J. R., *et al*. Common sense and nuclear peace. *Bulletin of the Atomic Scientists*, 39(4): 10s–13s, Apr. 1983, special supplement.

15. SAGLIO, J. F. Implantation des centrales nucleaires et l'environnement. *Annales des Mines*, pp. 139–44, Mar./Apr. 1976.

16. MARX, K., and ENGELS, F. *Selected works*. London, Lawrence and Wishart, 1970, pp. 349, 362, 364.

17. SANTOS, T. dos. The crisis of development theory and the problem of dependence in Latin America. In: BERNSTEIN, H. (ed.). *Underdevelopment and development: The Third World today*. Middlesex, England, Penguin, 1973, p. 61.

18. FURTADO, C. *O mito do desenvolvimento econômico*, 5th ed. São Paulo, Paz e Terra, 1981, p. 70.

19. FURTADO, C. Meio-ambiente, desenvolvimento e subdesenvolvimento na teoria econômica e no planejamento. In: ANDRADE, M. C., *et al*. *Meio-ambiente, desenvolvimento e subdesenvolvimento*. São Paulo, Hucitec, 1975, p. 83.

20. STRETTON, H. *Capitalism, socialism and the environment*. Cambridge, Cambridge University Press, 1976, pp. 48, 205–18.

21. FROMM, E. *To have or to be*. London, Sphere Books, 1979, pp. 172, 197.

22. WHITE, L., Jr. The historical roots of our ecological crisis. *Science*, 155(3767): 1203, Mar. 10, 1967.

23. LE GOFF, J. *La civilisation de l'occident médiéval*. Paris, Arthaud, 1967, p. 169.

24. HOLY BIBLE. O. T. *Genesis*, chap. 1, vss. 28–29. Catholic edition. New York, Nelson, 1966.

25. IOHANNES PAULUS II, P. P. *Redemptor hominis*. Rome, 1979.

26. BONO, E. *Ecologia e política à luz do Tao*. Porto Alegre, Brazil, Record, 1982.

27. RAO, B. R. Concepto de ecología en la literatura védica. *Mazingira*, 6(4): 65–77, 1982.

28. COMMISSION OF THE EUROPEAN COMMUNITIES. Continuation and implementation of a European Community policy and action programme on the environment. *Bulletin of the European Communities*, Supplement 6/76, pp. 50–51.

29. ILLICH, I. Outwitting the "developed" countries. In: BERNSTEIN, H. *Underdevelopment and development: The Third World today*. Middlesex, England, Penguin, 1973, pp. 357–68.

30. WYNNE, B. Redefining the issues of risk and public acceptance: The social viability of technology. *Futures*, pp. 13–31, Feb. 1983.

31. SCHUMACHER, E. T. *Small is beautiful*, p. 122.

32. BOLAFFI, G. A questão urbana, produção de habitações, construção civil e mercado de trabalho. *Novos Estudos Cebrap*, 2(1): 65, Apr. 1983.

33. RATTNER, H. Uma tecnologia para combater a pobreza. *Revista Brasileira de Tecnologia*, 12(2): 60–66, Apr./June 1981.

34. UNEP. *Directory of institutions and individuals active in environmentally sound and appropriate technologies*. Oxford, Pergamon Press, 1979.

35. BLANC, G. The world of appropriate technology. *Appropriate Technology*, 9(4): 14–15, Mar. 1983.

36. MEADOWS, D. H., *et al. Limits to growth*, chap. 4.

37. MADDOX, J. *The doomsday syndrome*. London, Macmillan, 1972, chap. 7.

38. FURTADO, C. *O Brasil pós-"milagre,"* 3d ed. São Paulo, Paz e Terra, 1981, p. 98.

39. DAGNINO, R. P. Indústria de armamentos: O estado e a tecnologia. *Revista Brasileira de Tecnologia*, 14(3): 6, May/June 1983.

40. UNEP. The environmental effects of military activity. In: UNEP. *The state of the environment, 1980: Selected topics*. Nairobi, UNEP, 1981, pp. 39–40.

41. CHOMSKY, N., and HERMAN, E. S. *The political economy of human rights*. London, Spokesman, 1979. Vol. 1: *The Washington connection and the Third World fascism*. Vol. 2: *After the cataclysm*.

42. KRANZBERG, M. Technology and human values. *Dialogue*, 11(4): 21–29, 1978.

43. FREIRE, P. *Educação e mudança*. São Paulo, Paz e Terra, 1981, pp. 22–23.

44. FREIRE, P. *Conscientização: Teoria e práctica da libertação; uma introdução ao pensamento de Paulo Freire*. São Paulo, Cortez & Moraes, 1980, p. 93.

45. QUIGLEY, C. Our ecological crisis. *Current History*, 59(347): 1–12, July 1970.

46. WESTING, A. H. A world in balance. *Environmental Conservation*, 8(3): 183, Autumn 1981.

47. PALME, O. The cost of over-kill. *Uniterra*, 6(5): 13, Sept./Oct. 1981.

48. EBERSTADT, N. Fertility declines in less-developed countries: Components and implications. *Environmental Conservation*, 8(3): 187–89, Autumn 1981.

49. EHRLICH, P. *Population bomb*. San Francisco, Ballantine, 1968.

50. BORGSTROM, G. *Too many*. New York, Macmillan, 1969.

51. MALTHUS, T. R. *Essay on the principle of population*. 7th ed. London, J. M. Dent, 1816.

52. PADDOCK, W., and PADDOCK, P. *Famine—1975!* Boston, Little, Brown, 1967.

53. EHRLICH, A. H., and EHRLICH, P. R. Dangers of uninformed optimism. *Environmental Conservation*, 8(3): 173–75, Autumn 1981.

54. WESTING, A. H. A world in balance, pp. 177–83.

55. BROWN, L. R., and SHAW, P. *Six steps to a sustainable society*. Washington, D.C., Worldwatch Institute, 1982, p. 13.

56. TAAGEPERA, R. People, skills and resources: An interaction model for world population growth. *Technological Forecasting and Social Change*, 13(1): 13–30, Jan. 1979.

57. COX, P. R., and PEEL, J. (eds.). *Population and pollution*. London, Academic Press, 1972.

58. VITTACHI, T. The world population: Back from the brink? *The Guardian*, p. 17, Aug. 28, 1979.

59. BELL, G., and TYRWHITT, J. *Human identity in the urban environment*. Middlesex, England, Penguin, 1972.

60. POPENOE, D. Urban sprawl: Some neglected sociological considerations. *Sociology and Social Research*, 63(2): 255–68, 1979.

61. UNEP. *The environment in 1982: Retrospect and prospect*. Nairobi, UNEP, 1982, p. 28 (UNEP/GC/SSC/2).

62. SCHMANDT, H. J., and BLOOMBERG, W., Jr. *The quality of urban life*. Beverly Hills, Calif., Sage, 1969.

63. CAMPBELL, A. *et al*. The quality of American life. New York, Russell Sage Foundation, 1976.

64. MICHELSON, W. *Environmental choice: Human behavior and residential satisfaction.* New York, Oxford University Press, 1977.

65. POPENOE, D. Urban residential differentiation. In: EFFROTT, M. P. (ed.). *The community approaches and applications.* New York, Free Press, 1974, pp. 35–56.

66. EL-SHAKHS, S. The population bomb and urbanization. *Ambio*, 12(2): 94–96, Apr. 1983.

67. PEREIRA, L. *Urbanização e subdesenvolvimento.* 4th ed. Rio de Janeiro, Zahar, 1979.

68. FARIA, V. Desenvolvimento, urbanização e mudanças na estrutura do emprego: A experiência brasileira dos últimos trinta anos. In: SORJ, B., et. al. *Sociedade e política no Brasil pós–64.* São Paulo, Brasiliense, 1983, pp. 124–42.

69. BOLAFFI, G. A questão urbana, p. 68.

70. ROBERTSON, J. Towards post-industrial liberation and reconstruction, pp. 6–24.

71. EHRLICH, P., and HARRIMAN, R. L. *How to be a survivor.* London, Ballantine, 1971.

72. EHRLICH, P., and EHRLICH, A. H. *Population, resources, environment: Issues in human ecology.* San Francisco, W. H. Freeman, 1970.

73. MESAROVIC, M., and PESTEL, E. *Mankind at the turning point: The second report to the Club of Rome.* New York, Dutton, 1974.

74. POQUET, G. The limits to global modelling. *International Social Science Journal*, 30(2): 284–300, 1978.

75. HECOX, W. E. Limits to growth revisited: Has the world modelling debate made any progress? *Environmental Affairs*, 5(1): 65–96, Winter 1976.

76. BISWAS, A. K. Estudios globales futuros: Una revision de la decada pasada. *Mazingira*, 6(1): 68–75, 1982.

77. U.S. COUNCIL ON ENVIRONMENTAL QUALITY. *The global 2000 report to the president: Entering the twenty-first century.* Washington, D.C., U.S. Government Printing Office, 1980.

78. FINSTERBUSCH, K. Consequences of increasing scarcity on affluent countries. *Technological Forecasting and Social Change*, 23: 59–73, 1983.

79. FUNDACIÓN BARILOCHE. *Modelo mundial latinoamericano: Informe presentado en el International Institute for Applied Systems Analysis, Vienna, Oct. 1974.* Bariloche, Fundación Bariloche, 1974.

80. BRANDT, W. From the Brandt Report. *IDR*, 4: 88–103, 1980.

81. GUERREIRO, R. S. A crise existe tanto no Sul como no Norte. *Jornal do Brasil*, Caderno Especial, p. 3, Nov. 1, 1981.

82. LEE, J. A. *Environmental security and global development: The essential connection.* Milwaukee, 1982.

83. IUCN. *World conservation strategy: Living resource conservation for sustainable development.* Gland, Switzerland, IUCN, 1980.

84. TALISAYON, S. D. New development goals and values in response to the global environmental crisis. *Science and Public Policy*, pp. 21–26, Feb. 1983.

85. BROWN, L. R. *Building a sustainable society.* New York, W. W. Norton, 1981.

86. BECKERMAN, W. *Two cheers for the affluent society.* London, St. Martin, 1976.

87. BECKERMAN, W. *In defence of economic growth.* London, J. Cape, 1974.

88. JACOBY, N. H. Organization for environmental management: National and transnational. *Management Science*, 19(10): 1138–50, June 1973.

89. CARSON, R. *Silent spring.* London, H. Hamilton, 1962.

90. LINERR, M. Gift of poison: The unacceptable face of development aid. *Ambio*, 11(1): 2–8, 1982.

91. ROTHSCHILD (LORD). *Risk.* London, BBC, 1978 (Richard Dimbleby Lecture, Nov. 1978).

92. ALTHUSSER, L. *Ideologia e aparelhos ideológicos de estado.* Lisboa, Presença, 1980, p. 77.

93. COMMONER, B. *Science and survival.* New York, Viking, 1967.

94. COMMONER, B. *Closing circle: The environmental crisis and its cure.* London, J. Cape, 1972.

95. DUBOS, R. *So human an animal.* New York, Scribner's, 1968.

96. EHRLICH, P., and HOLDREN, J. P. *Global ecology: Readings toward a rational strategy for man.* New York, Harcourt, Brace, Jovanovich, 1971.

97. NADER, R. *Ecotactics.* New York, Pocket Books, 1970.

98. STRETTON, H. *Capitalism, socialism and the environment.*

99. CASTRO, J. de. Subdesenvolvimento, causa primeira da poluição. *O Correio da UNESCO*, 1(3): 20, 1973.

100. KAHN, H. Our global growing pains. *Nation's Business*, pp. 32–38, July 1973.

101. GERELLI, E., and BARDE, J. P. *A qui profite l'environnement, riches ou pauvres?* Paris, Presses Universitaires de France, 1977.

102 ALMEIDA, M. O. The confrontation between problems of development and environment. *International Conciliation*, Jan. 1972, special issue: environment and development.

103. UNEP. Environment and development. *Facts*, UNEP FS/13, Mar. 1974, p. 2.

104. SACHS, I. Environment and styles of development. *Environment in Africa*, 1(1): 9–33, Dec. 1974.

105. SACHS, I. Meio-ambiente e desenvolvimento: Estrátegias de harmonização. In: ANDRADE, M. C., *et al. Meio-ambiente, desenvolvimento e subdesenvolvimento.* São Paulo, Hucitec, 1975, pp. 45–63.

106. SACHS, I. Sem medo de discordar. *Visão*, pp. 90–95, July 23, 1979.

107. WILKINSON, R. G. *Poverty and progress: An ecological perspective on economic development*. New York, Praeger, 1973.

108. MYERS, N. China's approach to environmental conservation. *Environmental Affairs*, 5(1): 33–63, Winter 1976.

109. WARD, B. *Progress for a small planet*. Middlesex, England, Penguin, 1979.

110. SAUVY, A. *Zero growth?* Oxford, B. Blackwell, 1975.

111. DORST, J. *Before nature dies*, p. 20.

2.

International Fora

Increases in oil prices, followed by a world energy crisis and economic recession during the seventies, have severely affected the Third World and demonstrated very clearly that sustained development and environmental balance depend on the interdependence of all nations. Within this context, environmental information plays an important role in providing national governments and supranational and regional organizations with elements for sound decisions.

Although the way each country deals with its own environment is likely to affect the whole world, not all nations have enough capital or technology to adopt the best measures for the use of natural resources. As a result of this deficiency, backward countries might disregard environmental safeguards in their efforts for development. Since all nations are conscious of this situation, management of the environment has become a subject about which concern is shared internationally, mainly as regards environmentally sound means of development.

To examine their own problems, various aspects concerning the environment have been discussed regionally by representatives of countries in the same continent. On a wider front, these national representatives also have discussed problems of the global environment under the sponsorship of both intergovernmental and international organizations. In both kinds of meetings, the approach of the debates has been political, economic or scientific, depending on the forum or the opportunity. Nevertheless, the way countries act in, and react to, the international discussion of environmental matters depends on the economic group they belong to: whether they are advanced countries or members of the Third World. The distinction is even more noticeable when voting on any proposition in international meetings (e.g., at the UN General Assembly): usually developing countries constitute one group and Western capitalist countries another, as their objectives are quite different. The communist countries will vote with either developing or developed countries, depending on the matter and their own interest. On the political side, the group of nonaligned countries is sensitive to Third World problems and positions since many developing countries belong to this group.

ORGANIZATIONS

The UN system, through its specialized organizations, agencies and regional commissions, has been one of the most active institutions in discussing the technological, economic and social aspects of the deterioration of the environment in developing countries. It has also led world efforts toward gathering and organizing data on global environment for international use.

To start with, the UN Secretariat-General itself has taken several environment-related initiatives on housing, transportation, science and technology. The General Assembly has always played an important part in all areas related to both development and environmental protection. In parallel, many UN divisions and specialized agencies formulate programs and projects concerning the environment in developing countries within their specific areas of action. The coordination, as far as the environment is concerned, has been centralized at the UN Environment Program (UNEP), which was established in December 1972 according to a recommendation of the UN Conference on the Human Environment.

Before that date, the catalytic function was performed by the Economic and Social Council (ECOSOC), which has been the intermediary between the UNEP and the General Assembly. The UNEP's role goes beyond the UN system, acting outside as a focal point for environmental action and then promoting coordination of efforts by governments and organizations around the world.

Along with this main goal, the UNEP has been seriously involved with demonstrating to the world that there exists a close relationship between the environment and development, and recommends that the United Nations should convene all possible means and ways to support financially and technically developing countries to achieve both progress and environmental benefits.

This idea has been brought to the practical plane through the UNEP's program, which concentrates on the following areas:

1. Human settlements and habitats
2. Human health and well-being
3. Land, water and desertification
4. Technology and the transfer of technology, trade and economics
5. Conservation of nature, wildlife, genetic resources and oceans
6. Energy

Two special groups of activities have been planned to give informational support to these areas of concentration:

- Earthwatch system of monitoring, information exchange and assessment, to be accessed by governments and world organizations; and
- Public information and education on environmental trends.

Ten years after its creation, the UNEP performed an indirect evaluation of its achievements by analyzing the state of the world environment, mainly in respect to the goals set for 1982.[1], [2], [3] The conclusions indicate that, even though problems are still very serious, international and regional actions to protect the environment have increased as much as people's concern for nature, and that the UNEP has played an important role in achieving that improvement. Corroborating these conclusions, its executive director postulated in his annual report of 1982 that, even if the UNEP had achieved only the reconciliation between development and environment, its existence would have been justified.[4] As a matter of fact, the last ten years showed that developing countries have had an ally in the UNEP, as it has always defended their right to development, based on the principle that "one of the single greatest threats to the environment is poverty."[5]

The UNEP carries on some activities by itself, such as promotion of conferences and meetings of experts, basic research to support specific joint projects or internal decisions and training of national teams of officers, research workers or experts. However, most of the activities are joint projects conducted with other organizations of the UN system, governments or scientific institutions.

The main collaborators for the environmental program within the UN system are:

1. The Regional Economic Commissions, which have sponsored development programs with a slant to the environment in areas such as water pollution, environmental health and the side effects of dams and power plants and have supported surveys to trace the state of the environment in specific regions, besides having performed informational and educational activities.

2. The UN Conference on Trade and Development (UNCTAD), which, as part of its more general interests and responsibilities in the area of international agreements on commodities, is concerned with the tariffs policy and any discriminatory measures imposed by developed countries, on the grounds of environmental considerations, upon the commodities exported or imported by developing countries.

3. The UN Industrial Development Organization (UNIDO), which has, as a general goal, given technical assistance to developing countries in the analysis of the environmental aspects of industrialization. It has been specifically involved in such projects as a survey of land-based sources of pollution of seas and a review of the environmental impact of all sources of energy and alternative styles of development and technology.

4. The UN Development Program (UNDP), which has supported different kinds of projects on environmentally sound development in the fields of industry, agriculture, housing and others directly related to the control of pollution and environmental sanitation.

5. The Food and Agriculture Organization (FAO), which has helped developing countries in the control of pests and diseases in crops, along with assessing environmental effects of agricultural chemicals on people and ecological systems and helping to avoid their undesirable effects. Supporting activities undertaken by FAO are training programs for experts, regional seminars and research studies.

6. The Intergovernmental Maritime Consultative Organization (IMCO), which is concerned with the harmful effects of chemicals and radioactive and various other wastes on marine biota, looking after the legal and scientific aspects of this particular type of pollution.

7. The International Atomic Energy Agency (IAEA), which has, as a fringe area of action, helped other organizations with research and effective control and prevention of radioactive contamination of the sea, fresh water, the atmosphere, land and crops. It has also advised them in the use of radiation techniques for environmental assessment and management.

8. The World Bank Group, consisting of the International Bank for Reconstruction and Development (IBRD), the International Development Asso-

ciation (IDA) and the International Finance Corporation (IFC), which has been financing the so-called development programs. In 1970, the position of environmental adviser was created in the World Bank and allocated to an expert. Later, this single position evolved into an Office of Environmental Affairs, which has taken environmental considerations into the formulation and appraisal of development projects submitted to the Bank for support. Among these projects are the construction of dams and roads, irrigation systems, sewage works, airports, power and fertilizer plants, petrochemical and mineral exploitation industries, information systems, education and health and welfare activities. Although the institutionalization of environmental advice for projects within a financial organization should be considered progress, comments at some national environmental agencies are that, due to the volume of projects, the advice has been based on a quite superficial analysis. Further, there are some reasons for developing countries to be still suspicious about the real role played by the Bank in this concern: as a matter of fact, would it not be a controller, representing interests of the transnational capitalism?[6]

9. The International Civil Aviation Organization (ICAO), which has participated in research for the establishment of standards and other actions related to noise in airports and generic noise pollution caused by planes.

10. The International Labor Organization (ILO), which is concerned with atmospheric pollution and other unhealthy conditions in places of work.

11. The UN Educational, Scientific and Cultural Organization (UNESCO), which is deeply involved in environmental programs as part of its scientific activities, such as Man and the Biosphere (MAB), Global Investigation of Pollution in the Maritime Environment (GIPME), international hydrologic program, multidisciplinary program of research and experiment on arid zones and humid tropic areas and many others.

12. The World Meteorological Organization (WMO), which has two main lines of research and activities related to environmental pollution: climatic changes caused by pollution and the abuse of nature and collection of world data to monitor background atmospheric pollution. Its program, World Weather Watch (WWW), operating through over 8,500 land stations and hundreds of ships, has been extended to include a network of air pollution monitoring stations.

13. The World Health Organization (WHO), which is involved with health aspects of environmental pollution, such as microbiological contamination of waters as a consequence of poor sanitary conditions and the effects of chemicals and fumes resulting from urban and industrial development. WHO created an Expert Committee on the Planning and Administration of National Programs for the control of adverse effects of pollutants. The Third World is quite confident in the benefits to come from WHO's program, "Health for All in the Year 2000."

14. The UN Advisory Committee on the Application of Science and Technology to Development (ACAST), which is the central body in the UN system dealing with the application of science and technology in the development process.

Outside the UN system many other international organizations are also dealing with environmental pollution, some of them touching indirectly on the interests of developing countries. Some of these organizations are:

1. The International Union for Conservation of Nature and Natural Resources (IUCN), which is a nongovernmental organization, founded in 1948. It has a scientific approach to the environment, focusing on the rational use of natural resources. To achieve this, IUCN supports research studies, promotes meetings of experts, prepares ecological guidelines for economic development of specific areas and publishes literature.
2. The International Council of Scientific Unions (ICSU), which develops research on environmental matters and established a Scientific Committee on Problems of the Environment (SCOPE).
3. The North Atlantic Treaty Organization (NATO), which is concerned with environmental threats in two ways: in the practical aspects related to combating the problem, by sponsoring pilot studies through the Committee on the Challenges of Modern Society (CCMS, created in 1969), and in its scientific aspects, by supporting research on environmental matters through NATO's Ecosciences Program, established in 1971. The main areas covered by the Ecosciences Program are sublethal toxicology, taxonomy, pollution indicator organisms and environmental data management. However, NATO's huge investments in war weapons bring concern to anyone who cares for the environment and for human beings.
4. The Council of Europe, which is acting in the field by promoting groups of study and conferences on pollution and by helping European countries in the development of their antipollution legislation. As part of this assistance, it has assigned a committee of experts to study existing legislation on pollution prevention; from this study resulted a Declaration of Principles ("Clean Air Charter"), intended to serve as a guide for member governments.
5. The Commission of the European Communities, which deals with pollution problems mainly through the Council of the European Communities' Working Party on the Environment. As part of its action program, the Commission, supported by the European Development Fund, declared interest in seeking common solutions to environmental problems with countries inside and outside the Community, including developing countries. However, it also

encourages polluting European industries to redeploy their plants into the Third World,[7] which constitutes a great concern for developing countries, since they will have to pay for a pollution they did not produce and for the profits they will certainly never benefit from!

6. The Organization for Economic Cooperation and Development (OECD), which has, among some others, two aims related to the subject in question: the highest possible economic growth and a rising standard of living in the member countries. Since 1970 it has had an Environment Committee, which is involved in urban environment, in water and air management and in controlling the undesirable occurrence of chemicals in the environment. Its *Interfutures*, projections into the world future, is considered a very balanced analysis of the global environmental situation.

7. The Inter-Parliamentary Union, which adopted, in its fifty-seventh Inter-Parliamentary Conference in 1969, a resolution on the role of parliaments in the protection of the human environment and conservation of natural resources.

8. The Organization of African Unity (OAU), which is concerned with African development and natural resources management. Some of its affiliated organizations are involved in research on animal health, soil and phytosanitation.

9. The Organization of American States (OAS), whose objectives include economic, social and cultural development of the Americas. As part of them, OAS is concerned, through the Inter-American Economic and Social Council, with better utilization of natural resources and improvement of the quality of life of people in the member countries.

As previously stated, the United Nations has been the most active international organization in relating environmental pollution problems to the crux of underdevelopment, looking for common solutions. Certainly, because the problems are big and varied, most of the components of the UN system work to some extent in this area. The great majority of the programs are joint efforts of UN organizations and are either promoted or coordinated by UNEP, the body that deals with environmental matters. However, many others are the initiative of different UN bodies and are not under direct coordination of UNEP. As a result, there is overlap of efforts in some areas, for instance, in marine pollution, causing dispersion of resources and information.

UNEP's support to outside projects is made possible mainly by the Environment Fund, whose resources are raised from member countries' contributions. This contribution has been decreasing in the last years, and world recession has been blamed, even by industrialized countries, for the decrease. But one could wonder whether developed countries' decreasing contributions were not caused by their disagreement with the UNEP's courageous position in defending the Third World's right to development and in unveiling the industrialized nations' policy of exporting polluting industries to developing countries.[8]

Information Activities of International Organizations

It is noticeable, particularly after the Stockholm conference, that supranational organizations have expressed interest in setting up one of two kinds of systems: either internal environmental information systems or international networks looking forward to monitoring and improving global environment conditions. In a few of these information systems the development component is also present.

The UN Information Systems. It seems that most of the efforts toward coordination or processing of environmental information for the use of the international community is made within the UN system. Three of these information systems are coordinated by UNEP, under the Earthwatch program: the International Referral System for Sources of Environmental Information (INFOTERRA), the Global Environmental Monitoring System (GEMS) and the International Register of Potentially Toxic Chemicals (IRPTC).

INFOTERRA is a network of international environmental information sources (individuals and organizations), locally coordinated by a focal point. In operation since 1977, by the end of 1982 this network involved 117 participating countries and a total of 9,500 sources. Since its establishment, INFOTERRA has been functioning as a referral system, but the team that concluded its evaluation in 1981 suggested that the system should progress toward the provision of substantive information as well; besides, it should enhance the use of the sources and improve the means of communication.[9]

INFOTERRA is closely linked to GEMS and IRPTC, which also work as environmental information sources. GEMS is the monitoring system of Earthwatch, designed to make an inventory of global resources and environmental quality. Operational since 1975, its activities fall into five major monitoring programs: renewable natural resources, climate, health, education, and the ocean. These programs have been developed in collaboration with specific UN bodies, such as FAO, WMO, WHO and UNESCO.[10]

The resistance GEMS sometimes meets from developing countries derives from the suspicion that, by monitoring the state of the environment (availability and quality of resources) within individual countries, GEMS might interfere in the internal affairs of the nations and endanger their sovereignty, as the system could progress toward becoming a tool for a supranational government.

The third information component of UNEP's Earthwatch is IRPTC, created in 1975 mainly to assist governments with sound information in order to reduce the hazards associated with chemicals in the environment. It functions in close cooperation with many international organizations, especially ILO and WHO. By the end of 1982, IRPTC's network had registered 104 national partners from ninety-five countries.[11]

Related to the above systems, UNEP also maintains an Industry and Environment Computerized Data Base, with technical information on pollution control.

Even though INFOTERRA, GEMS and IRPTC may be considered the main environmental information programs with global scope currently available, some other information units within the UN system also deserve being mentioned because they are of interest to the field of environmental pollution. These units are:

- The Marine Environment Data Information System (MEDI), cosponsored by UNESCO, Intergovernmental Oceanographic Commission (IOC), FAO, WMO and UNEP, is a referral system intended to provide information on the capabilities of existing centers supplying information on the marine environment around the world. Its operational foundations come from MEDI data bases, an inventory of marine data files compiled by IOC.
- The Soil Data Processing System, under the organization of FAO. Also under FAO is the Aquatic Science and Fisheries Information System (ASFIS), which delivers scientific and technological information on marine and fresh water environments.
- The International Reference Center for Waste Disposal, organized by WHO. Also organized by WHO are the Epidemiological Information System and the Appropriate Technology for Health Information System.
- The World Vigilance of Water Quality is a joint initiative of UNEP, WHO, UNESCO and WMO and serves as an information source for GEMS.
- The Global Investigation of Pollution in the Marine Environment (GIPME), sponsored by UNESCO and ICO, and Man and the Biosphere (MAB), sponsored by UNESCO, both of which actually generate and make accessible a

great deal of information concerning research conducted by them or on their behalf.

- (CIS) Centre International d'Information sur la Sécurité et Hygiène du Travail (International Center of Information about the Security and Hygiene of Labor) is ILO's information service on occupational safety.
- UNIDO, through its Industrial and Technological Information Bank (INTIB), provides information on technical alternatives available to developing countries for the preparation of industrial projects.
- The UN Disaster Relief Coordinator (UNDRO) operates a data bank on natural disasters.

To close the list, UN headquarters itself provides environmental information mainly through its Dag Hammarskjöld Library, UN Bibliographic Information System, Development Information System, Population Information Network and Statistical Office.

The European Communities Information Systems. In June 1971, the Council of Ministers of the European Community issued a resolution to coordinate the activities of its member states regarding scientific and technological information. The Council's attention was especially devoted to the creation and development of information systems toward the establishment of a European Network (EURONET). Later an action plan was prepared for the period 1975–77, and, as far as control of pollution was concerned, it foresaw an Environmental Management Information Network (EMIN) as part of EURONET. An Environmental Protection Information Group (EPIG) was created at the Committee of Scientific and Technical Information (CIDST) to coordinate efforts within the organization.

As a consequence of the plan of action and the recommendation of EPIG, many environmental information-related projects have been carried out by the Community and its member countries.

A specific source of information for EMIN is the Environmental Chemicals Data and Information Network (ECDIN), created in 1973 and developed by the European Community Joint Research Center, in Ispra, Italy. It is a hard-data system, which stores relevant information on any individual chemical compound for scientific and managerial use.[12] It is expected that closer links will be developed between ECDIN and UNEP's IRPTC in order to save global effort.

NGOs' Information Systems. Over 5,000 NGOs are known for their involvement in environmental activities, including those of education and information, in both developed and developing countries.

The efforts of a distinguished environmental NGO can be seen in the Environmental Law Information System (ELIS), designed and maintained by IUCN, with the assistance of UNEP and of the government of the Federal Republic of Germany. A computerized index of the endangered species mentioned in the legislation is also available at IUCN.

The Environmental Liaison Center (ELC), based in Nairobi, is another NGO of international status that deserves mention as an environmental information agency. ELC makes efforts to improve communication among NGOs, mainly in developing countries, and as a result, contributes to the Third World's pursuit of environmentally sound development. Many publications in the field, the organization of a directory on NGO environmental activities and the provision of the NGO Environmental Data System are part of ELC efforts to improve the information status of NGOs in developing countries.

UN INTERNATIONAL CONFERENCES ON THE ENVIRONMENT

In the beginning of the seventies, the environment started being discussed as a facet of development in UN conferences and at the General Assembly. As a result, the subject has often been approached from both political and economic points of view, rather than as a scientific and technological matter. The UN Conference on the Human Environment, held in Stockholm in 1972, is considered the turning point as regards the linking of environment and development. It was preceded and followed by other regional meetings of equal importance for developing countries. Besides these, the UNEP Governing Council has often discussed the environment-development relationship quite emphatically in its sessions, and the UN General Assembly is still the most important forum for the debate of this challenging subject.

From Founex into the Eighties

In 1971, a group of twenty-seven experts in the fields of development and environment met in Founex, Switzerland, at the UN's request, to discuss the interrelationship between these two fields and how developing countries could participate beneficially in the Stockholm Conference. The panel arrived at the conclusion that there are two broad categories of environmental problems: actual pollution problems caused by industrial activities and rapid urbanization, these problems being common to both developed and developing countries; and environmental disruption derived from poverty, a characteristic feature of underdeveloped countries.

Therefore, efforts toward an improved state of the global environment should, at the international level, include the creation of necessary conditions for the development of all nations. Within backward societies, betterment of the quality of life should aim at eliminating mass poverty and its worst forms of manifestation, such as malnutrition, squalor, disease and ignorance.[13]

Summing up, it could be said that this panel had two great achievements to its credit: broadening the concept of development to embrace social and political issues for people's benefit, including environmental interests; and broadening the concept of environment to incorporate human and social aspects, including development.

The conclusions of the panel resulted in the official position of developing countries at the UN Conference on the Human Environment, which took place in Stockholm, from June 5–16, 1972. Having taken the idea from the Founex panel that underdevelopment is in itself the worst kind of pollution, the Stockholm Conference looked at the environment in a broader way and from a social, human perspective:

· Of all things in the world, people are the most precious.
· The environmental movement could succeed only if there was a new commitment to liberation from the destructive forces of mass poverty, racial prejudice, economic injustice and the technologies of modern warfare.
· Environmental factors must be an integral part of development strategy.
· The concept of "no growth" could not be a viable policy for any society, but it was necessary to rethink the traditional concepts of the basic purposes of growth.
· The new technological order must be guided to achieve a better balance among the major elements which determine the level and quality of life.
· It is necessary to find new international means for better management of the world's common property resources and better means of exercising national sovereignties collectively with a greater sense of responsibility for the common good.[14]

The main achievements of the Conference were condensed in the "Declaration of the UN Conference on the Human Environment," which was then proclaimed as the core conclusions of the meeting and the common thought of the 113 nations taking part in it.

In the post-Stockholm period, the Symposium on Patterns of Resource Use, Environment and Development Strategies was an important event to channel the voices of developing countries in confirming their position about the relationship between environmental concern and development. This meeting was a joint effort of UNEP and UNCTAD and took place in Cocoyoc, Mexico, in October 1974. At the end of the Symposium, the basic conclusions and recommendations were adopted by the participants and issued under the title of *The Cocoyoc Declaration*.[15]

The above declaration emphasized that the root of environmental disruption lies in the socioeconomic structures within and among countries, and it also suggested that nations adopt a new system where human needs could be met without violating nature's limits.

This declaration also proposed that developing countries adopt alternative styles of development, more suitable to their local conditions and socioeconomic goals, according to the pattern of development freely elected by each country. As a part of this new strategy, ensuring the quality of life for all—present and future generations—was suggested as the main goal of development. All the proposals implied significant changes in the present patterns of growth, development and living standards in both developed and developing countries.

The seventies witnessed a high interest of the international community in discussing the problems connected with underdevelopment likely to affect environment. This interest still lasts in spite of world confrontation with many other problems.

Among the international meetings dealing with the above topics, some UN conferences deserve special mention, such as the World Conference on Population (1974), the second and third General Conference of UNIDO (1975, 1980), the Tripartite World Conference on Employment, Income Distribution and Social Progress and International Division of Labor (1976), the Conference on Technical Cooperation among Developing Countries (1978) and the Symposium on the Interrelations between Resources, Environment, Population and Development (1979). The content of discussion in these meetings showed evidence of the interconnection of problems pertaining to both the social and physical environments, the acuteness of their manifestation in the Third World and the common responsibility of nations to solve them through mutual cooperation and fair policies.

A meeting of specific interest for Latin America was organized jointly by UNEP and FAO in Bogotá, July 5–10, 1976. Latin American countries met then in order to discuss their regional problems related to the environment and development. They established priority lines of action and the administrative and legal structure that they would need. Similarly, many other regional meetings have since taken place in different parts of the world, mainly under the auspices of UNEP and the UN Economic Commission for each region.

In all the above mentioned meetings it implicitly or explicitly has been suggested that improvement of information services is part of the strategy to overcome underdevelopment and its associated environmental disruption. Emphasis is given to the international exchange of knowledge nd experiences, and it is especially suggested that developing countries should be granted free access to modern science and technology appropriate to their condition.

NOTES

1. UNEP. *Review of major achievements in the implementation of the Stockholm action plan*. Addendum: *Evaluation of the implementation of the 1982 goals*. Nairobi, 1981.

2. UNEP. *The state of the environment, 1972–1982*. Nairobi, 1982.

3. UNEP. *The environment in 1982: Retrospect and prospect*. Nairobi, 1982.

4. UNEP. *Annual report of the executive director, 1982*. Nairobi, 1983, p. 3.

5. UNEP. *Annual report*, p. 5.

6. GERRY, C. Restructuring underdevelopment? *Centre for Development Studies Newsletter*, 5: 3–7, Dec. 1981.

7. COMMISSION OF THE EUROPEAN COMMUNITIES. Continuation and implementation of a European Community policy and action programme on the environment. *Bulletin of the European Communities*, Supplement 6/76, pp. 50–51.

8. UNEP. *The state of the environment, 1981: Selected topics*. Nairobi, 1981, pp. 23–24.

9. MARTYN, J. *Report on the evaluation of INFOTERRA for the United Nations Environment Programme*. Paris, UNESCO, 1981, pp. 75–76.

10. UNEP. *Annual report*, pp. 31–35.

11. Ibid., pp. 37–38.

12. GEISS, F., and BOURDEAU, P. ECDIN, an EC data bank for envi-

ronmental chemicals. In: COULSTON, F., and KORTE, F. *Environmental quality and safety*. Stuttgart, G. Thieme, 1976, vol. 5, pp. 15–24.

13. UN. *Development and environment*. Report submitted by a panel of experts convened by the Secretary-General of the UN Conference on the Human Environment. Founex, Switzerland, UN, 1971, p. 25.

14. UN. *Report of the conference on the human environment*. New York, 1973, pp. 3, 45–48.

15. UNEP. *The Cocoyoc declaration*. Cocoyoc, Mexico, 1974.

3.

Environment in the Third World

A BLACK-AND-WHITE PICTURE

The environment in developing countries is said to be under threat, and this is certainly true: nature and humans are equally endangered creatures in those countries.

Lack of sanitation has polluted water and soil and jeopardized people's lives in all poor areas of the globe. It is estimated that 75 percent of the world's population lacks basic sanitation, and that 57 percent has no access to clean water.[1] This unhealthy situation has serious consequences for both the physical environment and population health, since WHO believes that 80 percent of disease cases—cholera, typhoid, hepatitis, leprosy, typhus, trachoma, schistosomiasis—in the Third World is caused by impure water and poor sanitation. As part of the same environmental problem connected with human settlements, housing conditions of seventeen Third World countries have been surveyed by researchers from the International Institute for Environment. They found that both housing and land tenure conditions of poor people are getting worse since Habitat.[2] An expressive example of this is seen in India and Nigeria, where over half of urban families live in one-room shelters.

In Asia, Africa and Latin America, landless peasants use the soil up to the point of its exhaustion and erosion. Another way people cause soil erosion and fertility loss is through clearance of forests, which happens mainly due to shifting cultivation or removal of wood for either energy purposes or grazing. GEMS monitoring data show that, from

1976 to 1980, closed forest removal yearly reached the amount of 1.3 million hectares in Africa, 4.1 million hectares in the Americas and 1.8 million hectares in Asia, with even worse projections for Latin America alone in the period 1981–85. Equally gloomy are the data concerning open forests in those continents.[3]

Erosion, loss of fertility and desertification are replacing forests in developing countries at the rate of 20 million hectares yearly;[4] in South America, Africa and Asia, 82 percent of the soils already presents serious limitations for agriculture.[5] This situation resulted in 450 million people—of whom 200 million are children—being chronically underfed in those places during the seventies,[6] and life expectancy at birth in developing countries is still 55.1 years.[7]

Poor education and lack of both economic and technological means for survival could explain the main causes of the Third World's environment degradation. However, this actually constitutes only one third of the picture.

A second cause for environmental damage are local government policies. Developing countries are given incentives to emulate patterns of development from industrialized countries, to look for a higher GNP and to spend more on technology than on satisfying human and social needs. An example of this bias is seen in the enormous investments made by poor countries like Brazil, Ecuador, Colombia, Nepal, Philippines, Pakistan, India and Egypt in building up gigantic hydrodams and manmade lakes. Receiving approval and loans from both developed countries and development aid agencies to fulfill their projects, those countries have cleared forests, destroyed genetic species, created conditions for the development of disease vectors and may have triggered unforeseeable conditions for global climate change. Besides these consequences, a financial commitment is contracted by those governments against society's will; nevertheless, society will have to pay for the debts, even though it was not given the chance to choose the kind of investment it would prefer.

Another aspect of the perverse economic policy of local governments (pressed by international development aid agencies and foreign banks) is the emphasis on the export of goods, even if, as a result, this action inflates internal prices and generates hunger. An example of such a policy can be observed first of all in Brazil, which exported all its 1982 production of maize and beans, which are the basic food of the low-income population. Another example is cow raising in Central

America and Colombia to feed North Americans. Another facet of the same environmental problem caused by overemphasizing export of local products are the cases of unfair exploitation of wood and mineral resources by multinational companies from Europe, the United States and Japan, which now replace the old colonizing metropolises.

The third factor of environmental degradation in the Third World is external, represented by policies of specific rich countries and international organizations. Developed countries usually impose severe restrictions concerning industrial pollution in order to clean up their environment and protect their people's health; thus, polluting factories are relocated in developing countries. For instance, the European Economic Community (EEC) suggests that European polluting industries be transferred to developing countries,[8] the Japanese aluminum industry has been relocated abroad, the petroleum refinery industry found a haven in Indonesia, and asbestos factories have been installed by the United States in Mexico and Brazil.[9]

Accepting this transfer of polluting industries would imply agreement with a new form of unfair international division of labor, where developed countries host "clean" research and industry, while the Third World hosts all the industrial pollution.

Brazil, as a typical developing country, has been producing and importing pollution, in spite of aspiring to a better environment. In this contradiction lies the core of its main environmental disruptions, which have been identified in interviews with environmental managers throughout the country:

- The problems concerned with the social environment are the most acute, mainly in relation to sanitary and housing conditions, public health, distribution of wealth and basic education.
- Fresh water pollution is a widespread problem, caused by domestic, industrial and agricultural waste, while atmospheric pollution affects only big industrial centers.
- All regions to a greater or lesser extent have been affected by land misuse problems associated with either agriculture or urbanization.
- There is a conflict between the central government's ambition for Brazil's short-term development and the ecological concern of both environmental agencies and the population.

Those problems still persist and have accrued by the magnified dimension of socioeconomic problems of the population, resulting from the economic policy long since adopted by the authoritarian government that has ruled Brazil from 1964 through a sequence of five military presidents.

Except for political control, which is not a feature of the Indian government anymore, the above environmental problems of Brazil may be compared to those of India, if one takes as reference the gloomy but humane view presented by the Center for Science and Environment in *The State of India's Environment, 1982*.[10] The main difference is that Brazil leans toward the industrialized world, leaving unsolved the problems of the hungry masses, while India is becoming viable through its own cultural model of development.

Although those problems have a local color and so demand a local approach, it is important to realize that many of those malaises seem to be common to most Third World countries. This can be inferred from the regional seminars organized by UNEP, in cooperation with UN regional economic commissions in 1979 and 1980, to analyze development patterns.[11] Some of those common environmental problems to be tackled are:

- Economic limitations, such as poverty, inequalities in the distribution of wealth, unemployment and underemployment;
- Health and sanitation insufficiency, such as endemic diseases, malnutrition and lack of both safe water supply and waste disposal facilities;
- Problems associated with urban and industrial development, such as rapid urbanization, poor housing conditions, various forms of pollution (industrial and agricultural), insufficient energy and transport inefficiency;
- Land-use malpractice, such as wildlife destruction, deforestation and related problems like erosion and desert encroachment;
- Consequences of adopting Western capitalist life and development style, such as overexploitation of resources by transnational companies, external dependence and loss of cultural identity.

Although innumerable projects of environmental interest are being implemented in most of the Third World countries individually—on their own initiative or that of UNEP or of any other international organization—the interchange of knowledge and experience among Third World countries to solve common problems is not yet the usual pattern. However, as their common problems have been identified, developing countries should now combine efforts in research, exchange of relevant information, organization of educational programs and various forms of technical cooperation in order to find their own model for sustainable development leading to appropriate solutions for their problems.

The Third World nations are not insensitive to the environment, and their concern is manifested both through governmental actions to prevent and to control pollution and through public awareness of ecological matters. In the following pages these manifestations will be analyzed with respect to environmental information, in aspects such as:

· Legislation as the basis of national environmental policy;
· The media's role in raising public consciousness;
· Educational programs leading both to the formation of ecologically minded citizens and to the advancement and communication of specialized knowledge.

Formal systems of environmental information will be dealt with separately.

GOVERNMENTAL ACTIONS

Governments of different countries, starting from the most technologically advanced, have responded to public manifestations against environmental nuisances by setting up a central environmental agency, by creating and enforcing legislation, and also by establishing quality standards, all equally important measures in the control of pollution. The first two of these measures have already been adopted by most of the developing countries. However, up to now only a minority has established its own standards, because this measure would require technological knowledge and a basic amount of equipment not possessed by the bulk of the Third World countries. Some others among them have just utilized alien models, but by adopting foreign standards they are frequently misled into imposing a too severe control upon emission, which is likely to discourage industrial expansion, so much welcomed by most of the developing countries. This contradiction suggests that some conflicts between developmental strategies and environmental policies still remain.

Usually, the first wave of legislation passed to prevent or to abate pollution in any country comes as part of a sanitary legislation package. This is why legislation about water pollution—the earliest to be developed as part of sanitation measures—is more voluminous and advanced. Examples of this early legislation on pollution are the United Kingdom Sanitation Acts of 1875 and the Public Health Act of 1848 and the U.S. Rivers and Harbors Refuse Acts of 1886 and 1899. Still, as a sanitary measure, the Brazilian Code of Waters, the earliest environmental legislation in the country, dates from 1934. This code has been later complemented by specific acts that regulate industrial pollution, such as the Edict SEMA 003/1975 (mercury control), Edict GM 013/1976 and Edict MINTER 0536/1976 (water quality criteria), Interministerial Edict 090/1978 (classification of watercourses), Edict MME 1832/1978 and Edict SEMA 002/1978 (industrial use of federal watercourses), Edict GM 323/1978 (water pollution by sugar mills) and the Edict MINTER 124/1980 (prevention of accidental water pollution by industries).

Nevertheless, the first comprehensive legislation on water pollution is North American, enforced in 1948. In the same line, Mexico adopted, in 1973, its Regulation to Prevent and Control Water Contamination, which was amended in 1975. India, in 1974, passed the Water (Prevention and Control of Pollution) Act, which is still in force.

The control of atmospheric pollution comes after that of clean water. England was the pioneer in air pollution control as a result of its earlier industrial development. In 1863, as part of the Alkali Act, the country got its first legislation about air pollution and industrial effluents. At the beginning of the fifties more strict control became necessary to abate London fog, and the Clean Air Act of 1956 was passed and later amended. Among developing countries it seems that India awoke earliest to the problem of air pollution, establishing its control through the Indian Motor Vehicle Act of 1939, via the adoption of a series of measures related to industrial pollution control (such as the Indian Boiler's Act of 1923) and finally through the Air (Prevention and Control of Pollution) Act of 1981.

The problems of solid waste disposal and noise abatement are quite recent in legislation. The United Kingdom Litter Act was the first legal control of solid waste in England, followed later by new legislation on the disposal of generic solid waste and specific waste, such as industrial and poisonous wastes. In the Third World, the central control of solid waste disposal is not yet frequently legislated under specific acts, but is integrated with either sanitary or industrial legislation. Nevertheless, the Brazilian government, through the Ministry of the Interior, passed the Solid Waste Edict no. 053 in 1979. In Egypt, as in other countries, solid waste is controlled at the municipal level.

English legislation about noise started with the Noise Abatement Act of 1960, one of the earliest in developed countries. In most developing countries, noise is either not yet considered a problem or its control is left to municipal authorities, as it is in the case of Egypt. Brazil, which adopted Edict MINTER 092/1980; India, with its Noise and Nuisance Act; and Mexico, that passed in 1976 the Federal Regulation for the Prevention and Control of Environmental Contamination by Noise, are the few exceptions having a national regulation, even though the control is loosely performed by local authorities.

Although very late in industrial development, many Third World countries have already adopted specific legislation to prevent and to control industrial pollution. This is the case in Mexico, whose first regulation dates from 1940. Brazil (Decree-Law 1413/1975, complemented by Decree 76.389/1975 and Decree 81.107/1977) and India (the Factories Act of 1948 and the Industries Development and Regulation Act of 1951) follow the same pattern.

Among developing countries the concern of the Indian government for the effects of some chemical products upon human health should be highlighted. This concern can be observed from India's adoption of the following regulations: the Poison Act of 1919, the Prevention of Food Adulteration Act of 1954, the Drugs (Amendment) Rules of 1964 and the Insecticides Act of 1914 (updated in 1968 and 1969). Egyptian Act no. 509/1954 also falls within the same orientation.

Another interesting facet of developing countries is the incorporation of environmental ethics in their political Constitutions. Examples of this incorporation are found in the Mexican Constitution (art. 27) and the Indian Constitution (arts. 48 and 51a), where the main framework for natural resources management is outlined.

Although developed and developing countries had started some kind of pollution control many years ago, a comprehensive view of environmental pollution is really an issue of the seventies, when those countries passed inclusive legislation and established their national environmental agencies, following recommendations of the Stockholm conference.

Among developed countries, the beginning of the seventies showed

West Germany (1970), France (1971), Japan (1971), England (1974) and the United States (1974) establishing their institutional structure for national environmental control.

As far as the Third World is concerned, W. Bassow, referring to a survey performed by the World Environment Center of New York, unveils that, by 1980, 102 developing countries already had some kind of environmental agency, in contrast with only 11 in 1972.[12] Some of those countries have created fully fledged ministries to deal with environment, as in Indonesia, Gabon, Ivory Coast, Upper Volta, Malaysia and Jordan. In other developing countries, environmental agencies have been embodied by either a secretariat, department, committee, bureau, board or council. Among the countries that have adopted this latter approach are Brazil, Mexico, Egypt, Ghana, Zaire, Uruguay, Oman, Iran, Thailand and Korea. India has a Department of Environment with the status of a ministry. Malaysia (1974), Iran (1974), Thailand (1975), Korea (1977), Indonesia (1981), Brazil (1981), Mexico (1982) and Oman (1982) constitute quite good examples of developing countries that have already approved comprehensive legislation on environment.

In a superficial analysis of the bulk of environmental legislation in the Third World—focusing mainly on Brazil, India, Mexico and Egypt—some facts show up: there are many gaps in the subject range covered; developmental policies conflict with environmental enactments; legislation toward natural resources conservation is quite voluminous; institutional control is sometimes duplicated because it is scattered under different authorities; and legislation neither keeps pace with national evolution nor reflects local culture. On the other hand, environmental authorities in developing countries recognize that legislation does not suffice, since they find it difficult to implement it because of lack of funds, technology and trained personnel.

Those drawbacks met by developing countries in the implementation of legislation result from characteristic underdevelopment scarcity which could easily be resolved if cooperation and interchange among nations were real and founded on a more just basis. In order to get all nations into the spirit of such a mutual assistance, the concept of "only one Earth" should also be reconsidered through an inward view: in spite of being artificially divided into countries, the Earth is a unity, essentially indivisible, if one considers vigilant concern for life as the guiding principle of international ethics.

PUBLIC AWARENESS

It is not possible to approach Third World advances toward environmental consciousness outside the international context since the ecological movement tends to be universal. Besides, in this concern, as in many other aspects, the Third World started later and has just followed the track of industrialized countries, sometimes being directly advised by foreign organizations or consultants.

The environmental movement presents two distinct phases according to the predominant interest in each period: first, the conservation of natural resources and, later, the struggle against pollution. From the end of the nineteenth century up to the early sixties, it was not really a popular movement, even in rich countries. There was concern by individuals and small groups, but the concern was directed only to the conservation of natural resources, including their protection and restoration. G. H. Siehl reminds us that most of the major conservation organizations active today were founded in the late 1800s and early 1900s, such as the Sierra Club, the Audubon Society, the U.S. National Parks Association, the Society of American Foresters and the International Association of Game, Fish and Conservation Commissioners.[13] Some other organizations, such as the Civilian Conservation Corps, Friends of the Earth and Resources for the Future were founded later. Besides their efforts in defending the environment, conservation organizations have played a major part in educating the public to the importance of protecting the environment, and quite certainly they may be acknowledged for preparing the actual environmental movement, which started in the 1960s.

The passage from a purely conservationist attitude to an antipollution cry is said to have been provoked by R. Carson's *Silent Spring*[14] in 1962. The emotional style she used to denounce the pollution of rivers and land by chemical pesticides alarmed the public, which increased its participation, both directly and through pressure on the government, to maintain vigilance on environmental issues. The danger of pollution and other environmental threats then became a popular subject among scientists, too.

The conservation organizations, which previously concentrated on the only objective of natural resources, broadened their interest to include the quality of the environment. Pressure groups have been created to fight specific problems. Different segments of the adult population have been sensitized. Youth expressed its concern about nature through the philosophy of the "hippie" movement. So, by the end of the sixties and early seventies the control of environmental pollution was already a sensitive subject with a very political content. This has happened in both developed and developing nations. Examples of NGOs undertaking the responsibility for public information and education in the developing world can be observed in India, where over 100 of such organizations are active in the field of the environment. In Indonesia there are 400 such organizations. The most traditional environmental NGO in Brazil is Fundação Brasileira para a Conservação da Natureza (FBCN). The Environment Liaison Center (ELC) in Nairobi has played a catalytic role in relation to the Third World NGOs, besides having taken some initiatives on its own in the fields of environmental education and information.

A particular case in Brazil is the strong presence of the Catholic Church in defending both the natural and the social environment. On the whole, the part played by the Church in the ecological movement consists of educating and alerting the population about threats to human rights, which includes the right to a better quality of life.

In some countries, like West Germany, England and France, the people have established an ecological party, which has been influential in government decisions. In Brazil, during the campaign for 1982 elections, some politicians characterized themselves as the "green candidates" and the most representative parties—PDS, the government party and PMDB, the main opposition party—included ecological concern in their platforms. The same movement was observed in the 1980 elections in India, when politicians incorporated environmental concerns in their programs as an answer to the the efforts of the Indian Department of Environment, which promoted the sensitization of politicians by publishing a book—*Environment: The Choice Before Us*.

In the United States, the Environmental Protection Act of 1970 provides for public participation in the protection of the environment. Many actions have since been taken by the EPA and similar agencies, as for instance, the Citizen Awareness/Information Exchange Program under the Federal Water Pollution Control Act Amendments (FWPCAA), the Inform and Involve Program (Forest Service), the creation of EPA's Office of Public Awareness and the different task forces incapsulated in many environmental programs.[15], [16]

In the same way, most governorates of Egypt have combined to create an advisory Environmental Committee, which includes members from both government and the public, the latter being represented by NGOs. This Committee advises local government in the identification and solution of problems that affect people.

In countries such as Indonesia and India, the national environmental agency, working with environmental study centers or different kinds of NGOs, stresses community awareness in their programs. In Egypt as well, the Ministry of Public Information is supposed to be preparing a program to heighten environmental consciousness in the population.

Although public awareness on environmental matters is a reality in every country, social scientists are still investigating the kind of participation people have in ecological movements, and they are also surveying people willing to participate. Thus, J. F. Wohlwill reviews previous research on the sociopolitical content of environmental attitudes.[17] The results of these previous research studies, plus J. F. Wohlwill's own findings and the observations made by W. W. Hines and G. E. Willeke,[18] indicate that environmental concern cuts across different segments of our society, although it has been found more strongly correlated with the following socioeconomic characteristics:

1. Income: the upper-middle class people, who tend to have higher income, are more likely to join the environmental movement, as long as it does not threaten their right to land and home ownership.
2. Education: a characteristic of the defenders of the environment is a good education, at either an average or a high level. They are also well informed through literature and the media.
3. Political preference: the environmental attitude has been associated with sociopolitical liberalism, for instance, the Democratic Party in the United States (as opposed to the Republican Party).
4. Social group: Ecological sympathy, as a manifestation of social group, is possibly more closely related to the dominant group than to racial minorities.

It seems that the above findings can be applied to developing countries as well. So, both C. E. Lins da Silva observing Brazil[19] and R. Claxton observing South America as a whole come equally to the conclusion that ecological concern in those places is restricted to elites in the affluent middle class.[20]

These comments could be extended to explain by analogy why the concern about the environment is an attitude more frequent in developed countries. In backward countries, natural resources are considered a commodity to be used in order to survive and possibly to reach development, their main goal. Where the protection of nature has to compete with alleviating hunger and oppression, it is understandable that the environmental movement is not so popular or it evolves slowly.

Considering the important role media can play in raising the population's ecological consciousness, some environmental agencies of developing countries—like Pakistan in 1979—have prepared special seminars and workshops on environmental awareness for journalists and other media personnel. Media have actually played an important part in educating and alerting the population against pollution risk and in voicing people's opinions and complaints about the quality of life they experience. It happens in both developed and developing countries, as concluded by C. E. Lins da Silva when analyzing North American and Brazilian contexts. He understands that newspapers have had an important political role in alerting the population to environmental issues, but he suggests that they should also catalyze society's efforts toward the attainment of solutions for environmental problems.[21] More specifically, W. W. Hines and G. E. Willeke show television as the most active medium in the environmental mission, followed by newspapers, the latter being important for people of higher education and social class.[18]

Television has proved to be an important medium to reach mass population through ecological messages in the Third World as well. In Brazil, from the late seventies on, news and documentaries on environmental topics have been shown on TV; even soap operas have assimilated the ecological motif. Mexican television (both governmental and commercial channels) presented in 1982 sixty programs on sanitary education prepared by the Secretaria de Salubridad y Asistencia, besides many others on the TV stations' own initiative. A daily ecological news bulletin is now programmed for TELEVISA (the private Mexican channel) in cooperation with the Subsecretaria de Mejoramiento del Ambiente. Preparation and distribution of short films on environmental topics for TV is also an important activity performed by the film divisions of the governments of India, Bangladesh, Philippines and Pakistan. Similar to TV is the power of radio among poor people. Thus, the *Voice of Kenya*, in collaboration with the Kenyan

National Secretariat for the Environment, presented a series of radio programs on various aspects of the environment. In Mexico, a daily radio program advises the population about community development. In Brazil even novels have discussed the environmental debate, as is the case of *Tieta do Agreste*, by the popular writer Jorge Amado, which approaches the pollution problem in Bahia.

J. W. Parlour and S. Schatzow investigated the role of the Canadian mass media (newspapers, magazines, TV and radio programs) in communicating news and in changing people's attitudes in relation to the physical environment from 1960 to 1972. These authors concluded that media played a very important part in that concern, mainly in the period between 1968 and 1971. Besides alerting the population, the Canadian media had the merit of getting the political recognition and the institutionalization of environmental concerns at all levels of the political system.[22]

Similarly, J. S. Bowman analyzed eight leading American mass circulation magazines from 1960 to 1979 and found that six out of eight devoted an increasing attention to ecological subjects. The author suggests that magazines could be the ideal media—in periodicity and style—to give a sustainable coverage to environmental issues.[23]

The power of media in changing governmental policies was apparent in India, when public pressure through the media caused authorities to reconsider projects like the Silent Valley hydroelectric plant in Kerala and the industrial complex of Mathura, which was likely to damage the Taj Mahal.

The argument against generation and use of nuclear energy, a constant theme of the media, was the subject of two important movies: in the cinema, through *The China Syndrome*, and on TV, through *The Day After*. On the musical scene, the suite *Minamata*, by Toshiko Akioshi, echoes complaints made by Japanese fishermen against the mercury poisoning of water in that town. In Brazilian popular music, a few good examples of ecological themes may be mentioned: *Águas de Março*, by Tom Jobim, *Baleias*, by Roberto Carlos, *Asa Branca*, by Luiz Gonzaga, and *O Reino Encantado da Natureza Contra O Rei do Mal*, the official tune of the Salgueiro samba school, written by Ivan Jorge for the 1979 carnival in Rio de Janeiro.

So varied are the problems and manifestations that, if one looks through newspapers or any publication that advertises public events in any metropolis of the world, one is certain to find either a demonstration or a meeting to protest against the armaments race, the installation of a nuclear power plant, the transport or burial of nuclear waste, the poisoning of water, land or air or some other kind of environmental threat.

All these observations show that, as the lay people become better informed about the causes and consequences of pollution, they begin to participate in the debate and are prepared to take action in defense of a better quality of human life compatible with respect for nature. That is why both environmental education and environmental information systems have such an important role in developing countries.

As far as environmental education is concerned, since 1976 and even more intensively after Tbilisi Conference, UNEP has been working with UNESCO to implement pilot projects in developing areas—Egypt, Kenya, Indonesia, India, Thailand and Latin America among others—aimed at either formal students or staff training. UNEP also helps governments in the preparation of regional seminars to orient teachers and school supervisors, as it did for Latin America in 1979. The same help is granted by UNESCO in the training of specialists, as it happened in Brazil, India, Argentina, Cuba, Philippines and Senegal.

Third World governments also have organized their own programs for both formal and informal education. So, some Asian countries—for example, Sri Lanka and Pakistan—have included environmental topics in the curriculum of primary schools. In central African countries, primary education, as well as secondary education, includes environmental matters, following the orientation of the African Curriculum Organization. Many Asian universities and research centers have introduced environmental education programs in their activities, for instance, in Thailand, Philippines and Indonesia. Also in the Philippines the important policy of teaching nonconventional energy technologies at secondary schools prevails. In Indonesia, priority is given to the specialization of university lecturers and government officials.

In Brazil, attempts have been made to introduce environmental education at primary and secondary levels, mainly by changing the focus of approach of conventional subjects. At Brazilian universities, environment is taught through either Master's courses (ecology, environmental engineering, and so on) or as a specific discipline. The Universidade de Brasília (UnB) has been the first one to introduce a Master's course in ecology. Environmental engineering is also part of all sanitary engineering courses. Workshops and seminars for the public are often prepared by universities. The Brazilian Secretariat of the Environment (SEMA) endeavours to assist all levels of environmental education and to prepare some special audio-visual materials as teaching help, mainly at primary and secondary levels. Most of SEMA's actions in this area—research, teachers' training, courses and so on—are performed in conjunction with state governments or other organizations.[24]

The Mexican government considers education as fundamental for the success of environmental programs. That is why the Secretaria de Educación Publica decided to integrate ecological concepts in the existing programs and disciplines at all levels of formal education, always taking into account the real economic, social and cultural needs of the country. According to this approach at both primary and secondary levels, environmental education is emphasized in disciplines connected with natural and social sciences, technological and health education, discussed from the viewpoint of the community interest.

In higher education, the broad objective of Mexican policy is still related to forming positive ecological attitudes in future professionals in both social and technological careers. However, at this level, specific courses of environmental interest exist at the Universidad Nacional Autónoma de México (UNAM) (environmental engineering, hydrological engineering, biochemical engineering, chemical engineering and marine sciences), at the Universidad Autónoma Metropolitana (ecology, environmental engineering, biology and chemistry) and at the Instituto Politécnico Nacional.

At the Master's level, the environment constitutes a specialization in four Mexican universities, while at the Universidad Nacional Autónoma de México environment is the object of a Ph.D. program.[25, 26]

Regarding informal education, the Subsecretaria de Mejoramiento del Ambiente established and is implementing a program of ecological training divided into three phases: emotional appeal, enhancement of concern and information for self-reliance of the whole population.

In India, emphasis is first of all given to primary education, where students analyze their own living environment and the importance of nature conservation to human beings. In order to select topics, to guide teachers in the use of appropriate teaching methods and to prepare educational material (books, audio-visual materials, special kits and so on), the Department of Environment works closely with the National

Council of Educational Research and Training. At a higher level, many universities have already created courses on environmental sciences. Jawaharlal Nehru University (JNU), entirely funded by Indian government, maintains courses at the postgraduate level at the School of Environmental Sciences, taught by an interdisciplinary group of high-standard professors and researchers, who have behind them the support of a good library, macro- and microcomputers for bibliographic and scientific research and well-equipped laboratories.

In Egypt, policy and orientation for environmental education are provided for by the National Board of Environmental Education, part of the National Academy of Scientific Research and Technology. The general system of education, at the basic level, offers students short lessons about their living environment, and at the university level, students are encouraged to take disciplines of environmental content. There are ten universities in Egypt and all of them have at least one course on sanitary engineering and another on public health medicine. At the High Institute of Public Health in Alexandria, both the Environmental Health Department and the Occupational Health Department offer diploma, Master's and Ph.D. courses. Research at Egyptian universities is mainly funded by sources from the United States, where most of the scientists are also trained. Germany has also cooperated by offering equipment.

From the above described panorama of environmental education in the Third World—especially from what happens in Brazil, Mexico, India and Egypt—one realizes that most of the burden of formal education lies on governmental structure and funds; some foreign cooperation exists for research support and high level qualification of manpower only; NGOs are deeply involved in the work concerning informal education of the population; and industries are not willing to cooperate with either government or universities for the promotion of environmental education and research.

Information and education together have the potential to lead the population to care for nature, since they are the basis for public awareness and for changes in attitudes. This is the reason to believe that all the above actions concerning environmental education and information indicate that developing nations—governments and society—are progressively becoming conscious of their environmental problems. On the other hand, thanks to the media, NGOs and even to official environmental agencies, the population in those countries have slowly realized they have duties and rights in regard to a clean environment. In spite of the remaining economic and sociopolitical hindrances, environmentalism within Third World societies evolves into an enduring concern, and this move cannot be reversed.

In addition to this process of the environmental awakening of the Third World, it has been evidenced that information is a strategic element for nations to reach environmentally sound development. But, like a vicious circle, the existence of an environmental information system in developing countries is determined by factors inherent in the characteristics of underdevelopment, as discussed in the next chapter.

NOTES

1. UN. *International drinking water supply and sanitation decade: Report of the Secretary-General*. New York, 1980.

2. HARDOY, J., and SATTERTHWAITE, D. *Shelter: Need and response; Land and settlement policies in 17 Third World nations*. New York, J. Wiley, 1981.

3. UNEP. *Annual report of the executive director, 1982*. Nairobi, 1983, pp. 32–33.

4. EL-HINNAWI, E. El medio ambiente mundial: ¿Ahora, hacia donde? *Mazingira*, 6(1): 62, 1982.

5. TOLBA, M. The challenge of the eighties. *Uniterra*, 6(1): 11, Jan./Feb. 1981.

6. UNEP. *The environment in 1982: Retrospect and prospect*. Nairobi, 1982, p. 29.

7. Ibid., p. 28.

8. COMMISSION OF THE EUROPEAN COMMUNITIES. Continuation and implementation of a European Community policy and action programme on the environment. *Bulletin of the European Communities*, Supplement 6/76, pp. 50–51.

9. UNEP. *The state of the environment, 1981*. Nairobi, 1981.

10. CENTER FOR SCIENCE AND ENVIRONMENT. *The state of India's environment, 1982: A citizens' report*. New Delhi, 1982.

11. UNEP. *Choosing the options: Alternative lifestyles and development patterns*. Nairobi, 1980, pp. 7–10, 48–49, 53–59, 70–76.

12. BASSOW, W. Third World is "going environmental." *The Journal of Commerce and Commercial Banking*, New York, Dec. 30, 1980, p. 4.

13. SIEHL, G. H. Our world—and welcome to it! *Library Journal*, 95: 1443–47, Apr. 15, 1970.

14. CARSON, R. *Silent spring*. London, H. Hamilton, 1962.

15. NICHOLSON, J. M. Citizens can have a voice in environment decision-making. *Catalyst for Environment Energy*, 7(2): 28–31, 1980.

16. POGELL, S. M. Government-initiated public participation in environmental decisions. *Environmental Comment*, 4: 4–6, Apr. 1979.

17. WOHLWILL, J. F. The social and political matrix of environmental attitudes and analysis of the vote on the California Coastal Zone Regulation Act. *Environment and Behavior*, 11(1): 71–85, Mar. 1979.

18. HINES, W. W., and WILLEKE, G. E. Public perceptions of water quality in a metropolitan area. *Water Resources Bulletin*, 10(4): 745–55, Aug. 1974.

19. LINS DA SILVA, C. E. Jornalismo e ecologia. *Comunicação e Sociedade*, 7: 54–55, Mar. 1982.

20. CLAXTON, R. Ambientalismo latino-americano: Fraco e limitado às elites. *Raízes*, n. 2, Aug. 1977. *Apud*: LINS DA SILVA, C. E. Jornalismo e ecologia, p. 54.

21. LINS DA SILVA, C. E. Jornalismo e ecologia, pp. 51–61.

22. PARLOUR, J. W., and SCHATZOW, S. The mass media and public concern for environmental problems in Canada, 1960–1972. *International Journal of Environmental Studies*, 13(1): 9–17, 1978.

23. BOWMAN, J. S. Environmental coverage in the mass media: A longitudinal study. *International Journal of Environmental Studies*, 18: 11–22, 1981.

24. BRAZIL. Secretaria Especial do Meio Ambiente. *Meio ambiente no Brasil: Evolução e perspectiva histórica*. Brasília, 1982, pp. 76–78.

25. MEXICO. Comisión Intersecretarial de Saneamiento Ambiental. *México: Diez años después de Estocolmo*. Nairobi, 1982, pp. 67–73.

26. MEXICO. Subsecretaria de Mejoramiento del Ambiente. *Informe nacional gobierno de Mexico*. Nairobi, PNUMA, 1982, pp. 20–22.

27. The author is also indebted to the following individuals and institutions, from or through which she got important sources of information or specific data included in this chapter.

International Panorama
    International organizations—UN Center for Human Settlements (Kenya); Pan-American Health Organization (United States and Mexico); UNEP (Kenya and Mexico); UN Dag Hammarskjöld Library (United States).
    Individual experts on the international environment—Mr. Y. Ahmad (UNEP); Mr. P. Bartelmus (UN Statistical Office); Mr. P. Bifani (UNEP); Mr. U. Dabholkar (UNEP); Dr. R. Goodland (The World Bank); Mr. K. Grose (UNEP Library); Dr. A. Khosla (UNEP); Mr. T. Munetic (UNEP); Mr. A. C. Printz, Jr. (USAID); Mr. H. Sakimura (UNEP).
Brazil
    Organization—Secretaria Especial de Meio Ambiente (SEMA).

Individual experts—Dr. Paulo Nogueira Neto (SEMA); Mr. Estanislau Monteiro de Oliveira (SEMA).

Egypt

Organizations—Academy of Scientific Research and Technology; General Office for Building, Housing and Planning Research; General Organization for Physical Planning; General Organization for Sewerage and Disposal Control.

Individual experts—Dr. G. El-Samra (National Committee for the Protection of the Environment); Eng. Kamel Maksoud (General Organization for Industrialization—GOFI); Dr. Fatma Gohary (Water Pollution Control Lab—National Research Center); Dr. Ahmed Hamza (High Institute of Public Health—Univ. of Alexandria); Dr. Samia G. Saad (High Institute of Public Health—Univ. of Alexandria); Mr. David Spiller (The British Council—Cairo).

India

Organization—Department of Environment (DOE).

Individual experts—Mr. A. Agarwal (Center for Science and Environment); Dr. D. Banerjee (JNU—Center of Social Medicine and Community Health); Mr. P. R. Banerjee (DOE/INFOTERRA); Mr. Bhardwaej (DOE); Dr. D. K. Biswas (DOE); Prof. J. M. Dave (JNU—School of Environmental Sciences); Dr. A. Lahiri (Department of Science and Technology); Mr. H. S. Matharu (Central Board for the Prevention and Control of Water Pollution); Dr. T. Mathew (Environment Service Group, The World Wildlife Fund—India); Mr. M. K. Moitra (Ministry of Works and Housing); Mr. M. Parabrahman (DOE); Mr. M. C. Swany (DOE); Dr. V. Venugopalan (Ministry of Works and Housing).

Mexico

Organizations—UN Economic Commission for Latin America (ECLA); Secretaria de Agricultura y Recursos Hidraulicos (SARH); Subsecretaria de Mejoramiento del Ambiente de la Secretaria de Salubridade y Asistencia (SMA).

Individual experts—Eng. S. Agis (SMA); Eng. Alcineaga (SMA); Eng. F. Bahamonte (SARH); Mr. J. Hurtubia (UNEP Reg. Office); Mr. F. Lopez de Alba (SAHOP); Dr. F. Mascareño Sauceda (SMA); Eng. L. Sanchez de Carmona (SAHOP).

4.

The Design of Environmental Information Systems in Developing Countries

Economy and sociology of development are still evolving and there is neither a conclusive definition nor a universally accepted interpretation of underdevelopment, as one can deduce from R. H. Chilcote's review.[1] Having this argument in mind and avoiding playing the expert, this chapter attempts to identify—from living experience rather than from theory—some key variables to understand the connections between underdevelopment and environmental information.

Taking this personal approach to underdevelopment, two groups of political and socioeconomic factors, as presented in Table 1, can strongly influence the design, implementation and management of government environmental information systems in developing nations. These determinants derive partially from the very nature of society of backward countries, that is, from their underdevelopment. There are some basic features of underdevelopment—focused on in the next section—which are common to all developing countries and likely to interfere with the quality and sometimes with the very existence of an official environmental information system. Some of those features (for example, bureaucratization) are not peculiar to underdeveloped countries. However, in combination with the other factors of underdevelopment they are more likely to be a drawback to the existence of information systems than when they occur in advanced countries.

The other group of intervening factors that will be discussed in this chapter results from the environmental policy adopted by a government in the management of its natural resources and in the control of pollution. Three basic kinds of policies of environmental ideologies can

Table 1 : FACTORS AFFECTING THE DESIGN OF ENVIRONMENTAL INFORMATION SYSTEMS IN DEVELOPING COUNTRIES

Features of Underdevelopment	Components of Environmental Policy
1. Poverty	1. Strategy on natural resources
2. Dualism	2. Industrial development objectives
3. Technological inadequacy	3. Pollution criteria and parameters
4. Unproductive bureaucracies	4. International exchange platform
5. Contingent nature of development programmes	5. Attitude towards public participation
6. Control of information media	6. Subject areas of R & D encouraged
7. Dependency	
8. Nationalism	
9. Deficient educational system	

be adopted: conservationist, ecodevelopmental or technocratic. Each of these lines considers the environment from a different perspective, as discussed in chapter 1. The focus and concern officially given to the management of the environment will determine the level of priority given to an environmental information system. Since the users of such a system would be principally authorities in environmental agencies, governmental policy would be a strong indicator of the type of demands for data and services placed upon the information system.

The identification of some of these two groups of factors has been tried, and the result is exhibited concisely in Table 1. In this respect, the factors discussed hereafter should be seen not as overriding influences but as tendencies that in any case will confirm the sensitive character of an environmental information system and its political role in a developing country.

IMPLICATIONS OF UNDERDEVELOPMENT

The status and characteristics of underdevelopment affect every single area of planning, including that of information systems.

H. Bernstein sees development as resulting from a desire to overcome malnutrition, poverty and disease and embracing such features of social justice as equality of opportunity, full employment, generally available social services, equitable distribution of income and basic political freedoms. H. Bernstein comments that this scenario embodies a value judgment concerning society's needs,[2] which suggests the principle of self-determination as a right of developing countries.

His statement is an indirect description of the syndrome of underdevelopment, which the poor nations strive to overcome. Some indication of these problems can be gained by examining in Table 1 the features of underdevelopment, which are relevant factors in the life cycle of an environmental information system.

The first factor—poverty—can be expressed both through shortage of capital as far as government finance is concerned and paucity of per capita output. According to P. Baran, these are the two main characteristics of backward countries and account for their designation as underdeveloped.[3]

As far as information systems are concerned, the main implication of poverty, in both its facets, is that information is not a priority at all. Poor countries have to struggle for survival, and their scarce resources are allocated to basic areas, such as health, agriculture, education, transportation and the implementation of industries. As a consequence, the entire information process, from its generation to the internal dissemination and exchange among the agencies, is seen as a luxury and either kept at an inadequate level—quantitatively or qualitatively—or not embarked upon at all. Collecting and processing local data are problematical and unsatisfactory as the result of deficiencies in equipment and personnel. Attempts to access foreign information meets technological barriers (like obsolete computers or inadequate communication systems) and financial restrictions (such as devaluation of local currency at the international financial market, balance of payments, reimbursement of royalties and strict financial legislation concerning importation). Last, salaries are low and do not attract professionals from the environmental sciences to work as information scientists.

The second characteristic of underdevelopment is what the structuralists call a "dualistic economy," resulting from the penetration of modern capitalistic enterprises into archaic structures. According to C. Furtado, it is likely to produce a hybrid structure, part tending to behave as a capitalist system, part perpetuating the features of the previously existing system.[4]

It is worth mentioning that the concept of dualism is rejected by authors like L. Pereira[5] and A. G. Frank,[6] who prefer to explain underdevelopment through either a historic or a more dynamic approach. Disregarding theoretical discrepancies among specialists, dualism has been included in the present chapter with the intention of emphasizing the contrasting faces of the socioeconomic reality of developing countries. These faces correspond to different stages of the history (either evolution or involution) of different regions or cultural groups, which coexist with unequal shares of benefits and opportunities within the same country. Regardless of terminology, this idea expresses the level of development characterized by L. C. Bresser Pereira as "industrialised underdevelopment."[7] The exogen and endogen causes of such inequalities, as well as strategies for their elimination, are important matters still waiting for practical solutions.

Duality is reflected on the information scene by the fact that there are often just two kinds of services: either the "grass roots" information of a practical character for some specific groups of workers or the scientific and technological information for the elites. An example of the first is the dissemination of agricultural information for rural workers, usually conducted at the instigation of, and with help from, FAO. The second—scientific and technological information for research workers—is more prestigious and receives a better share of scarce resources. This happens partially because of its elitist characteristics and links with the international scientific community. A further reason is that, in the eyes of government, information for research workers indirectly supports technological advancement, a goal usually set in development programs. Unfortunately, most of the scientific and technological research follows the lines of developed countries in order to satisfy demands of the higher layers of society in developing nations. These higher layers of society have a pattern of consumption that resembles that of developed countries, which is a feature of dependency, as discussed by C. Furtado.[8] Thus, society as a whole does not benefit from scientific and technological research and local science and technology do not reach their autonomy.

Information support from government is often given to the scientific community by supplying them with either foreign ''models'' (patents, processes and so on) or theoretical information to tackle specific problems on a short-term basis. This narrow focus may be explained by the existing tendency in developing countries not to invest in long-term research programs, usually because they are pressured by immediate needs.

This leads to the consideration of the third factor, which is the existence in developing countries of a technological inadequacy or inviability (called technological ''gap'' by the analysts from developed countries). Lack of resources, information and freedom cause developing countries to select and to use a technology that is not often the most convenient to their own needs or to their local environment. This impasse leads developing countries to a kind of technological inviability, followed by perplexity upon realizing that local problems need local solutions. To make this situation even more complex, poverty and inefficacy of communication interfere in the exchange with other countries within the Third World. This inadequacy sometimes affects the process of information transfer, noticeably in electronics (data processing and communication), transport and micrographics.

Sharing information resources among organizations within the same country is a big problem where long distances cannot be overcome by good communication and transportation systems. This same barrier is found when trying to access international networks or database systems if the appropriate communication technology is not available.

Poor countries face with concern the prospect of a paperless society, like that foreseen by F. Lancaster,[9] as they are not technologically prepared to participate in this progress. Far behind in being self-sufficient, they have no money to import the necessary technology from abroad, and yet they could not survive without information generated in the advanced countries. New technology invented by developed countries for their own practical purposes and economic reasons creates enormous problems for developing countries.

One of the options presented to the Third World is the adoption of alternative technologies more suitable to their present conditions. However, as seen by D. S. Price,[10] the adoption of intermediate technologies would probably mean a return to the traditional printing stage as the West goes paperless, a return to mechanical processes while the developed countries take advantage of electronics and holography. Obviously, these options very often do not satisfy developing countries either. It is sometimes believed that intermediate technologies are merely methods promoted by developed countries to divert the more backward countries from full development. The adoption of alternative information technology would not solve the problem, however, if foreign information was presented in a higher technological form because developing countries would not be able to integrate with an international network to "decode" the information they could get. The question of scientific information and the technology necessary to its processing in developing countries has therefore the same socioeconomic and political components as the whole subject of development itself.

P. Baran points out the existence of a vast unproductive bureaucracy and a military establishment as typical handicaps of backward countries, and that they are among the principal obstacles to their economic growth.[11] When developing countries are ruled by a military government, a frequent situation, the result is military red tape in official organizations and services. Just as this factor obstructs general development, so it also negatively influences the promotion of information systems:

• A huge amount of resources is wasted in these redundant institutions to the detriment of the productive sectors of the country.

- An excessive hierarchization slows the administration and the decision-making process hinders communication between different levels within the same organization and among different ones. A further problem for information systems is that, in such governments, they are usually placed at a low hierarchical level.
- Military beliefs and attitudes promoted by the central government make it very rigid and illiberal. Because defense is overemphasized by a military government, excess secrecy obstructs the free flow of information.
- An internal struggle for power among civil organizations and between civil and military organizations results in gaps and overlaps in the areas to be covered, in loss of authority and in discontinuance of the system or service in question. As far as information is concerned, there is a tendency for some "fashionable" subareas to be covered by a number of institutions while others receive no attention at all.

Linked to the above is the fact that development programs are, in practice, contingent upon a number of other factors. As a result of these contingencies, some government programs never come to be reality; others are interrupted, with a waste of effort and resources. Both internal and foreign pressures are responsible for these contingencies along with incompetence of authorities, absence of long-range planning and lack of resources. In such conditions, the stability of an environmental information system will depend on:

- The convictions of the ruling group in the central government about the value of formal information;
- The position given to information in general and to the specific subject—environment—in the main plan of government and government's short-range goals toward development;
- How highly assessed is this specific sector supported by the information system, both by the internal and international market; and
- Approval granted by foreign creditors and political controllers.

The sixth factor listed in Table 1—control of information media—is a part of the political scenario within which the movement to development occurs. All means to bring information and education to the population are under certain kinds of control by both national government and foreign bodies, the latter being represented by foreign government agencies (including intelligence services), transnational companies, political groups and supranational organizations. National government itself operates through a system of either direct or indirect censorship, since it is usually very sensitive to criticism, does not believe in the public's capacity for participation and values excessive secrecy about official plans and data.

Government secrecy affects the transfer of information to the local population and to foreign governments. Not only are newspapers, radio and television affected, but the printing industry also is affected, mainly as concerns publishing in the area of the social sciences (books and periodicals). In addition to censorship, government may control the media by imposing severe economic constraints, like limitation of credit and obstacles to importing equipment and material. Both forms of control upon the media are highlighted by P. Marconi, who uncovers the history of the Brazilian press in the period 1968–78,[12] and by W. A. Rugh, who discusses the Arab press.[13]

Foreign influence in the control of information channels is part of a broader politico-economic domination of the poor by wealthy countries. So, it is manifested through both politico-ideological control and economic measures. Control may be exercised over the government, a specific national group or economic area or directly over the population. Thus, funding may be offered to a project or withdrawn from it according to the circumstance, depending on the interest of the specific foreign group or country at the time. Political ideology is disseminated in reading material and conferences on the grounds of foreign cooperation in the national program of education. Public opinion is frequently misled by campaigns carried through the media by foreign economic groups interested in starting a business in the country, which would despoil the environment as a result.

Another facet of the international control of information channels is represented by the international news agencies. Rare and powerless are these agencies in developing countries; only from 1978 on have some of them joined the Yugoslavian Tanjub in a pool of nonaligned countries. So, news is transmitted to the world via international agencies (AP, AFP, UPI, Reuters), which tend to select maliciously the news to be published or distort the message received from local correspondents. This unfair situation has been the object of attention from UNESCO, which produced the MacBride Report, defending a new international order of communication and information.[14] A more direct analysis of the transnational news agencies is presented by F. Reyes Matta.[15]

Government control over environmental information may become even tighter because, overall, the subject area deals with the knowledge of existing natural resources, the performance of industries and the health condition of the population. National governments try to hide this kind of information from the local public to avoid pressures and from other governments in order to preserve national sovereignty. Foreign governments, however, do their best to obtain this key information so that they may achieve domination of the political scene and have a hold on the economic potential of the observed country. Despite the vigilance of developing countries, data on their natural environment can be collected and processed with more accuracy by developed countries, thanks to the higher technology owned by the latter; for instance, images obtained by satellites, which are easily available in the United States, but sometimes considered a "national security matter" in the country from which the image is taken.

Dependency on developed countries is another characteristic of underdevelopment and is apparent, in a broader context, in the slavish imitation of foreign models by developing countries. This attitude is contradictory, since they also reject intellectual colonization and overvalue their own national characteristics. This is the result both of media control by foreign agencies and of a whole politico-economic manipulation of backward countries by the developed nations, which has prevailed for centuries. Foreign products and ideas are promoted so massively in developing countries that they inhibit the generation of a national model or an appropriate technology.

R. Prebish blames transnational companies for this situation, as they promote the intnernationalization of consumption in the Third World, causing new forms of dependency and deeper inequalities within those societies.[16] T. dos Santos identifies this syndrome with underdevelopment itself and calls it "dependent capitalism,"[17, 18] while A. G. Frank explains it through his theory of "metropolis-satellite polarization."[19, 20] Both theories imply a condition of stagnation, within which the dominated country is tamed to consume products and information that the dominant country wants to disseminate and to sell. In such a context, an information system in an underdeveloped country must attempt to deal with (and may break) the monopoly of information by some foreign groups and with the difficulty in selecting what is really needed from what is an artificial need created by the external producer or controller.

It is a further complication that, within this setting, an environmental information system has the role of encouraging and supporting research groups to develop indigenous knowledge and consequent technology appropriate to the local environment.

Reaction to foreign domination can range from "idolatry" of foreign models to an overestimation of national values. Nationalism is a common feature of developing countries, which is revealed at the international forum by overstressing each country's right to self-determination and control over its own natural resources. Extreme nationalism tends to have a negative effect on information systems, as it makes the country impermeable to external ideas and developments, besides creating obstacles to the exchange of information. Here again, the question of secrecy is relevant, since oversecrecy might impede the supply of information about specific aspects of the environment, such as natural resources and health, to both foreign countries and supranational organizations.

A last factor—a deficient educational system—is one of the basic handicaps of developing countries. As far as environmental information systems are concerned, it affects all levels: information professionals need better training in both information techniques and subject content; the specific users, that is, the environmental managers and scientists, are not aware of the information resources available and how to obtain and use them; and the whole population needs to have free access to all levels of education and information in order to be able to achieve a better quality of life and then value a clean, balanced environment.

IMPLICATIONS OF ENVIRONMENTAL POLICY

The sensitive nature of a national environmental information system makes it entirely dependent on the policy adopted by the central government in relation to the human and the natural environment, since it is intended to support government programs, and it is, moreover, maintained by government.

Six points have been identified in Table 1 as the main components of environmental policy which could determine the basic characteristics of an environmental information system. These components are:

1. Strategy on natural resources
2. Industrial development objectives

3. Pollution criteria and parameters
4. International exchange platform
5. Attitudes toward public participation
6. Subject areas of R&D encouraged

Conservationist Policy

When the central government adopts a conservationist policy in the management of environment the results may be as follows:

Strategy on natural resources. Emphasis will be on natural resources, in a search for their protection for ecological reasons and their preservation for future generations. As a whole, in this situation there will be a smaller volume and less variety of data to be managed by the information system than under either of the other two policies. However, the area of natural resources will probably be the strongest feature in the system.

Industrial development objectives. The government will probably not encourage the implementation of industries because it values a clean environment more than the economic growth of the country. Environmentally sound technology will be welcome, even though government is not deeply involved in supporting its development. Thus, the information system will have a weak monitoring activity and lack of data on pollution and the socioeconomic environment but a good deal of information on alternative technologies.

Pollution criteria and parameters. Regulations to control pollution are very restrictive as part of the scheme to discourage industrial development. There will be emphasis on the promotion and collection of studies like environmental impact statements.

International exchange platform. Conservationists are usually prone to considering the environment as a common inheritance of all living beings. If held in governmental circles, this attitude is likely to lead authorities to adopt, in diplomatic and international discussions, an open position about environmental subjects. It may also lead to a disposition toward the free international exchange of information and willing participation in any international information networks.

Attitude toward public participation. Government will welcome public participation in environmental affairs. However, the fact that government is already meeting environmental needs may mean that less public pressure is exerted on the subject and, accordingly, that less information support is demanded. If it really occurs, the information system will not be a priority for government unless channelled for public education on ecological matters.

Subject areas of R&D encouraged. Government will tend to sponsor educational and research programs in subject areas related to natural sciences and landscape planning, and so the supporting information systems will be sponsored as well. There will be a tendency for users of the information system to be scientists and researchers, followed by environmental managers.

Ecodevelopmental Policy

When government has taken an ecodevelopmental line, the whole planning system is directed to environmentally sound socioeconomic development, presenting the following specific tendencies:

Strategy on natural resources. Nature is seen as a benefit to be shared by the whole world and a resource to be thoughtfully used and individually managed by each country. Information will be needed about: (a) the basic components of the environment (air, water, soil), both from the theoretical approach to nature and from a practical view of local environment; (b) products of soil and water, the so-called renewable natural resources, from both naturalistic and economic points of view and under both global and local approaches; (c) mineral resources, their nature, reserves, careful means of exploitation and research data on potential renewable substitutes; and (d) recycling of resources.

Industrial development objectives. Industrial expansion is an important issue, and government emphasizes the development of technology appropriate to the physical and the socioeconomic environment of the country. Transfer of technology is considered in accordance with priorities of the recipient country, and exchange of information—especially on AT—among developing countries is highly considered. The information system will develop a collection of data and documents on technological and socioeconomic facets of industrial development. Hard data will be demanded as much as qualitative information. All possible information on resources—conserving, raw material saving, low-energy, low-waste and nonencroaching technologies should be collected and made available. Information should also be supplied on methods of assessing environmental impact and undesirable deterioration by natural disasters and human misuse. The social subset of the information system dealing with the human environment is going to be highly considered and demanded by the government.

Pollution criteria and parameters. Pollution control regulation is quite comprehensive and more flexible, as the parameters vary regionally according to the carrying capacity of the local environment. This flexibility is part of the strategy to attract industries and keep them under the control of the local environmental agency. Monitoring systems, in this case, must be quite complex, and the resulting data are going to be a subset of the information system. The environmental impact statements are equally valuable. Information on the social aspects of pollution associated with poverty and bad sanitary conditions might receive high consideration.

International exchange platform. Ecodevelopment pays particular attention to developing countries sharing the same problems and similar environmental conditions. However, most of the studies and research are carried out in developed countries. Backward nations adopting an ecodevelopmental policy are usually open to an international exchange of information, even though they are also very zealous in the pursuit of their national sovereignty. Developing countries will always expect to receive information from the developed world, but they tend to be reluctant to supply information about their own resources and policies if the requesting country is not an ally or does not belong to the same politico-economic group. In this context, the developing country may participate officially in international information networks, but the ambivalent nature of its participation is likely to create some practical difficulties with the other members and ethical problems for information workers.

Attitudes toward public participation. Government will encourage public participation, which is expected to occur in all representative groups of society, which will be manifesting different positions. It will result in high pressure on the government and, consequently, on the information system. Data and document collection must range from conservation of natural resources and the social aspects of the human environment to high technology for industrial development, not disregarding information to the general public about means of solving routine environmental problems. All forms of community information will come under the umbrella of an ecodevelopment policy of government.

Subject areas of R&D encouraged. Government will give incentives to educational and research programs in a variety of environment-related subjects, such as natural sciences, agriculture, environmental and sanitary engineering, AT, environmental education, environmental medicine and urban and landscape planning. Therefore, environmental information systems must be prepared to satisfy demands related to these programs. Users will also represent the following variety of interests: environmental managers, governmental planners, scientists and technologists, social scientists, researchers, lecturers and, by extension, the population as a whole.

Technocratic Policy

A technocratic government tends to mean development at any cost, with environmental disruption accepted as a by-product of progress. In these circumstances the tendencies are as follows:

Strategy on natural resources. Nature is regarded solely as an economic resource to be used in the most profitable way. The information required about natural resources will always show the economic bias of government and will stress the demand of quantitative aspects and the available technology to exploit existing resources. Data on natural resources, mainly mineral reserves, are a fundamental requirement of government and usually considered a sensitive subject.

Industrial development objectives. A technocratic government tends to understand development as being synonymous with economic growth and to consider industrial expansion as the way to reach it. It will have little concern for protecting the environment and preventing pollution. Information will be demanded on cheap, profitable technologies, whether or not they are harmful to the environment or to people.

Pollution criteria and parameters. Pollution control regulation is quite limited and parameters very lax, so that industries are not deterred from coming to the country in question. Some monitoring is performed in order to alleviate the consquences of heavy pollution. Social information related to pollution of poverty may eventually receive some consideration, with marginal interest, to back up development programs.

International exchange platform. Backward countries with great ambitions to develop try to get the maximum in resources and knowledge from outside in order to reach their goals more quickly. At the same time they attach some secrecy to information about their own natural reserves, which they plan to exploit on their own behalf in the future. These countries have a very conflicting international relationship, as they need and want help, but they are not willing to renounce some rights to their national sovereignty in exchange. It rebounds on the information scene by creating an ambiguous position: central government controls main decisions about international exchange and is eager to receive information but is very secretive in giving it to foreign countries. This position makes it more difficult for them to participate in international networks or to collaborate with information systems of supranational organizations.

Attitude toward public participation. Government will not give incentives to public movements in favor of protecting the environment, but these will certainly occur as a result of the official neglect of the natural and the social environment. Nor will programs of environmental education be given much incentive. However, information systems will be asked to follow the development of the ecological movement, at both national and international levels, in order to provide government with necessary information to support its policies and promote its good environmental public image. The information system will probably be a high priority for a technocratic government, as it associates scientific and technological information with economic development.

Subject areas of R&D encouraged. Government will give incentives to development of applied knowledge. R&D information will be accordingly promoted, mainly in areas of industrial technology, agriculture and all different branches of engineering, where the abatement of pollution will also be included. Users will be mainly government planners and technologists.

Since the political process is very dynamic, the overall policy adopted by government in the management of a country's environment will determine the specific policy of environmental information. Reciprocally, that policy is also likely to be influenced by the information available to policymakers. In the potential of information professionals to persuade environmental policymakers lies a great deal of their power to cooperate for the betterment of life quality on Earth and especially in the Third World. This improvement cannot be achieved without environmentally sound development and the assurance of justice for all.

NOTES

1. CHILCOTE, R. H. Teorias reformistas e revolucionárias de desenvolvimento e subdesenvolvimento. *Revista de Economia Política*, 3(3): 103–23, July/Sept. 1983.

2. BERNSTEIN, H. (ed.). *Underdevelopment and development: The Third World today*. Middlesex, England, Penguin, 1973, pp. 13–14.

3. BARAN, P. A. *The political economy of growth*. Middlesex, England, Penguin, 1973, pp. 26–27.

4. FURTADO, C. Elements of a theory of underdevelopment—The underdeveloped structures. In: BERNSTEIN, H. (ed.). *Underdevelopment and development*, p. 34.

5. PEREIRA, L. *Ensaios de sociologia do desenvolvimento*. São Paulo, Pioneira, 1970, pp. 57–58.

6. FRANK, A. G. Desenvolvimento do subdesenvolvimento latino-americano. In: PEREIRA, L. *Urbanização e subdesenvolvimento*. 4th ed. Rio de Janeiro, Zahar, 1979, pp. 26–27.

7. BRESSER PEREIRA, L. C. Semiverdades e falsas idéias sobre o Brasil. *Novos Estudos Cebrap*, 2(2): 24, July 1983.

8. FURTADO, C. *O mito do desenvolvimento econômico*. 5th ed. Rio de Janeiro, Paz e Terra, 1981, chap. 2.

9. LANCASTER, F. W. *Towards paperless information systems*. New York, Academic Press, 1978.

10. PRICE, D. S. Some aspects of "world brain" notions. In: KOCHEN, M. (ed.). *Information for action: From knowledge to wisdom*. New York, Academic Press, 1975, pp. 177–92.

11. BARAN, P. A. *The political economy of growth*, pp. 376–77.

12. MARCONI, P. *A censura política na imprensa Brasileira, (1968–1978)*. São Paulo, Global, 1980.

13. RUGH, W. A. *The Arab press: News media and political process in the Arab world*. Syracuse, N.Y., Syracuse University Press, 1979.

14. UNESCO. *Un solo mundo, voces múltiples: Communicación e información en nuestro tiempo*. Mexico City, Fondo de Cultura, 1980.

15. REYES MATTA, F. *A informação na nova ordem internacional*. Rio de Janeiro, Paz e Terra, 1980.

16. PREBISH, R. A crise do capitalismo maduro. *Novos Estudos Cebrap*, 2(1): 22, Apr. 1983.

17. SANTOS, T. dos. The crisis of development theory and the problem of dependence in Latin America. In: BERNSTEIN, H. (ed.). *Underdevelopment and development*, p. 76.

18. SANTOS, T. dos. *Imperialismo y dependencia*. Mexico City, Era, 1978.

19. FRANK, A. G. *Capitalism and underdevelopment in Latin America: Historical studies of Chile and Brazil*. Middlesex, England, Penguin, 1969, pp. 32–36.

20. FRANK, A. G. Desenvolvimento do subdesenvolvimento latino-americano, pp. 25–38.

5.

Environmental Information Systems in the Third World

Quite understandable from the technological point of view, a great deal of information about the environment of the Third World is processed and available firsthand in developed countries, as for instance, through GEMS, WWW or, in the case of satellite images, in the USA. However, this information covers only a part of the subject and only from a single-sided perspective. To span the whole picture of the Third World environment, this information should be associated with local information available through formal systems of environmental information or through scattered nonconventional sources of information, such as nonpublished reports, government officers, journalists and environmental experts. Some of the institutional sources may be found using INFOTERRA as a starting point. Facing such a mosaic, information about the Third World environment becomes an intriguing puzzle, which always demands sagacious efforts from the individual researcher to integrate and to validate data.

Another problem causing the international community to be poorly informed about the Third World environment and its management is the shortage of publications that take a global approach to the subject or that discuss socioeconomic implications of environmental management within the context and from the viewpoint of underdevelopment. For different reasons, both foreign and local writers fail to give a clear picture to the public. Even the few foreign scientists and journalists attentive to environmental issues in the Third World tend to analyze them from the standpoint fashionable at a given time as, for instance, at present, the conurbations of Latin America and Asia or the deforestation of the Amazon region. Very frequently these analyses are flawed because of the lack of comprehensive knowledge of the situation and the politico-economic bias of Western authors. On the other hand, local writers usually approach environmental themes from the specialized perspective of their individual education, for example, sanitary engineering, medicine, botany, geology and so on. Despite the existence of some integrated understanding of environment, theoretical knowledge is still compartmentalized, partly because the inclusion of this holistic view in university curricula is still incipient in developing countries. In some countries, like Brazil, Chile, Libya, Iran and Zaire, local authors meet a governmental barrier to their freedom of expression, mainly concerning those aspects of the environment considered to be ''matters of national security,'' such as natural resources, public health, war potential and human rights.

Having in mind the above picture and the fragile sociological structure of environmental information systems in the Third World as discussed in chapter 4, one can be guided through the labyrinth of local specialists, institutions and documents by identifying some key organizations and experts in the country or at least in the overall region. If searching from outside, the specific national embassy may be a good referral source (in spite of the usual bureaucracy and consequent delay) to be used in conjunction with the INFOTERRA directory.

As one may infer from the above picture, environmental information in developing countries is a subject involved in a web of technological, economic and political problems still to be resolved, and both national and international communities have their parts to play in this context for the common benefit of humankind.

LATIN AMERICA

As far as the present status of environmental information systems is concerned, in the Third World it seems that Latin America is ahead in the institutionalization of information services. First of all, the UNEP Regional Office for Latin America (ROLA), located in Mexico City, gives support and incentive to several institutions in the production, organization and dissemination of environmental information in Latin America and the Caribbean. It has worked as a referral center and as an agent of public information. Complementing ROLA's action, the Institute for Renewable Natural Resources Development (INDERENA) in Bogotá, Colombia, is INFOTERRA's model coordinator center for Latin America and is prepared to refer researchers to regional sources of environmental information.

Closely associated with the UNEP Regional Office, the UN Economic Commission for Latin America (ECLA) is equipped to give overall information about the socioeconomic environment of the region and even about its natural environment. ECLA, on different occasions, has promoted seminars on environmental themes and also has undertaken in-depth research on the Latin American environment in its ecological, political, economic and social dimensions. An NGO interested in disseminating information on appropriate technology and science for development is the Centro de Estudios Economicos y Sociales del Tercer Mundo, located in Mexico. Finally, the Centro Internacional de Formación en Ciencias Ambientales (CIFCA), located in Madrid, whose aim is environmental education of Latin Americans and other Spanish-speaking people, could also be consulted successfully.

There are also some information systems covering specific environmental aspects, such as:

- Planning, urban and regional development: LATINAH, the Latin American Information Network on Human Settlements, in Bogotá, Colombia, sponsored by the Centro Nacional de Estudios de la Construción and the Universidad Nacional de Colombia; INFOPLAN, the Information System for Planning in Latin America and the Caribbean, in Santiago, Chile, sponsored by ECLA's Centro Latinoamericano de Documentación Económica y Social (CLADES); and SINDU, the Inter-American Information Service on Urban Development, in Bogotá, Colombia.
- Health and sanitation: REPIDISCA, the Pan-American Network for Information and Documentation on Sanitary Engineering and Environmental Sciences, maintained by the Centro Panamericano de Información y Documentación en Ingeniería Sanitária y Ciencias Ambientales (CEPIS), in Lima, Peru; the documentation center of the Asociación Interamericana de Ingeniería Sanitaria y Ambiental, in Rio de Janeiro, Brazil; BIREME, the Regional Library of Medicine and Health Sciences, in São Paulo, Brazil; and ECO, the information system of the Centro Panamericano de Ecología Humana y Salud of the Pan-American Health Organization (PAHO), in Mexico. Both REPIDISCA and BIREME have support from PAHO as well.
- Food and agriculture (including soil recovery, use of pesticides and fertilizers): AGRINTER, the Inter-American System of Agricultural Information, cosponsored by the Instituto Interamericano de Cooperación para la Agricultura (IICA), in San José, Costa Rica, and FAO; the documentation center of

the Colombian Centro Interamericano de Agricultura Tropical (CIAT); Empresa Brasileira de Pesquisa Agropecuária (EMPRAPA) in Brazil; and Centro Nacional de Informação Documental Agrícola (CENAGRI) in Brazil.

- Population: DOCPAL, the Documentation System on Population in Latin America, sponsored by ECLA's Centro Latinoamericano de Demografia (CELADE) in Santiago, Chile. DOCPAL integrates POPIN, the Population International Network, and has an associated system (DOCPOP) in São Paulo, Brazil.

A number of organizations in Latin America are involved in research and information dissemination on appropriate technology and ecodevelopment: Centro de Ecodesarrollo in Mexico, Centro Mesoamericano de Estudios sobre Tecnología Apropiada (CEMAT) in Guatemala, Clearinghouse of Peace Corps Intermediate Technology in Nicaragua, INDERENA in Colombia, PLANALSUCAR of the Instituto do Açucar e do Álcool in Brazil, Instituto de Pesquisa Tecnológica (IPT) in Brazil, Secretaria de Asentamientos Humanos y Obras Publicas (SAHOP) in Mexico, SELAVIT, the Latin American and Asian Grassroots Housing Service in Chile, Servício de Documentación y Comunicación Rural (SEDOC) in Nicaragua, and Servício de Información Técnica in Ecuador.

Besides the above regional information systems and local organizations, each country possesses its own internal information structure. At least two more of the organizations can be helpful for those who are interested in the environment of Latin American countries: the national environmental agency (for example, SEMA in Brazil) and the national council responsible for scientific and technological development of a specific country (for example, the Brazilian Council of Scientific and Technological Development (CNPq) and CONACYT, the Mexican Council of Science and Technology).

Brazil

Regarding a comprehensive approach to the Brazilian environment (present status, problems and actions toward solutions), two kinds of sources of information deserve being mentioned: national bibliographies and environment-related institutions (as a referral source for documentation and baseline data). The development of both sources depends on political factors. Thus, local information through the traditional means of monographs and magazine articles is scarce in Brazil because of some components of the present picture of the country:

- The federal government has adopted a technocratic policy, which gives the environment little attention.
- The environmental system of the government lacks resources and is still trying to systematize its knowledge of the whole situation under the coordination of one body.
- The study of environmental sciences at the university level is quite recent. Thus, there are few experts in the country capable of a holistic approach to the subject.
- The printing industry is poor. This situation forces publishers to concentrate scarce resources on publications that could have a larger public. The potential number of consumers is already limited by illiteracy and poverty. The environment is not yet a candidate for a best-seller in Brazil.
- The central government is not seriously interested in the promotion of mass education on the concern for environment, since public information and participation in official programs is not a feature of a dictatorship (sometimes called "relative democracy" by members of the central government), the form of government in power in Brazil since 1964. As a result, environmental matters have only recently become a subject for debate among a broader audience, thanks mainly to the media. One can even foresee that democracy, meaning freedom of expression and public participation, will soon be reintroduced in Brazil via the ecological movement.
- Censorship, which was explicitly imposed through 1979 by the central government and controlled by the military, prevented printing and the media's capacity to report on some governmental policies and facts, such as the ones related to the management of the country's natural resources and public health.

In spite of these handicaps, there are some good sources of information on the available Brazilian environmental literature, for instance:

- Centro de Informações em Ciências Ambientais (CICA) of the Universidade Federal do Rio Grande do Sul, which published, in 1983–84, the two initial volumes of *Bibliografia Brasileira de Ciências Ambientais*. This work constitutes part of a data base, which is to be publicly available soon.
- Instituto Brasileiro de Informação em Ciência e Tecnologia (IBICT), under CNPq, which responds for the national policy of scientific and technological information. It publishes the national union catalog of periodicals. It has also published *Fontes de Informação em Meio Ambiente no Brasil*, a guide to

Brazilian sources of environmental information, which lists the main Brazilian periodicals dealing with the subject.

- Centro de Informações Nucleares (CIN) of the Comissão Nacional de Energia Nuclear (CNEN), which gives input to International Nuclear Information
  System (INIS) and maintains its own data base of national publications on
  nuclear energy as well. A new data base, FONTE, has now been made
  available by CIN, and constitutes bibliographic references on alternative sources
  of energy.
- CENPES, Centro de Pesquisa e Desenvolvimento Leopoldo Miguez A. de
  Mello, the main information and documentation center of Petróleo Brasileiro
  S. A. (PETROBRÁS), which has just opened to the public the data base
  SINAL on conventional and nonconventional literature about alternative sources
  of energy.
- SIJUR, which is the information system managed by PRODASEN, the data
  processing unit of the Brazilian Senate. The information service offered by
  SIJUR relies on a data base that includes legislation, books and periodicals.
  Quite a large network of organizations on legal matters is now connected on
  line to SIJUR.
- SEMA, which manages its internal system SIMA for information about the
  literature available at its library (SRD) and about legislation (SRL). SEMA
  is also INFOTERRA's focal point in Brazil.

To find out about environmental organizations, a few directories can
be used:

- The already mentioned *Fontes de Informação em Meio Ambiente no Brasil*,
  published in 1983 by IBICT, in collaboration with SEMA, which also includes environmental management institutions and documentation centers.
- The Unidade Referencial (UNIR), a data base of referral sources on science
  and technology, being prepared by IBICT.
- *Catálogo Nacional das Instituições que Atuam na Área do Meio Ambiente*,
  a national directory of environmental organizations (governmental and otherwise), whose updated version was published in 1982 by SEMA, the national agency of environmental control.
- *Catálogo Brasileiro de Engenharia Sanitária e Ambiental*, a national guide
  of governmental organizaions, NGOs, consultants, labor associations, on-going
  projects and so on in the areas of environmental and sanitary engineering,
  published yearly by the Associação Brasileira de Engenharia Sanitária e Ambiental (ABES).
- *Habitat: Guia de Bibliotecas e Centros de Documentação*, a guide to Brazilian libraries and documentation centers specializing in urban development.
  It was published by CNPq's Coordenação de Saúde, Nutrição e Habitação,
  with 1981 data.

Among the large number of environmental organizations in the country, some of them can work as access points in specific sub-areas:

• Scientific and technological development: SICTEX of the Ministry of Foreign Affairs, an information service available at the main Brazilian embassies of the world; IBICT, in Brasília, through its documentation center on science policy (CPO); and CNPq's sectoral boards (Brasília).
• Natural resources: SEMA (Brasília); Superintendência de Recursos Naturais e Meio Ambiente (SUPREN) of Fundação Instituto Brasileiro de Geografia e Estatística (FIBGE) in Rio de Janeiro; Departamento Nacional de Produção Mineral (DNPM), under the Ministry of Mines and Energy (Brasília), mainly through its RADAMBRASIL project of natural resources survey and its information center in Rio de Janeiro; Instituto Brasileiro de Desenvolvimento Florestal (IBDF) in Brasília; and Fundação Brasileira para Conservação da Natureza (FBCN) in Rio de Janeiro.
• Pollution control: SEMA, in Brasília; SUPREN, in Rio de Janeiro; Departamento Nacional de Águas e Energia Elétrica (DNAEE), under the Ministry of Mines and Energy (Brasília); Companhia de Tecnologia de Saneamento Ambiental (CETESB) in São Paulo; Fundação Estadual de Engenharia do Meio Ambiente (FEEMA) in Rio de Janeiro; Centro de Pesquisas e Desenvolvimento (CEPED) in Salvador; and Secretaria de Tecnologia Industrial (STI) in Brasília).
• Public health and sanitation: Fundação Osvaldo Cruz (Rio de Janeiro); Departamento Nacional de Obras e Saneamento (DNOS) in Rio de Janeiro; Divisão Nacional de Biologia Humana e Saúde Ambiental, under the Ministry of Health (Brasília).
• Alternative technology: Instituto Nacional de Tecnologia (INT) in Rio de Janeiro and Secretaria de Tecnologia Industrial (STI) in Brasília, both under the Ministry of Industry and Commerce (MIC); CEPED (Salvador); PLANALSUCAR (Piracicaba); and Fundação Centro Tecnológico de Minas Gerais (CETEC) in Belo Horizonte. All these organizations are willing to transfer appropriate technology to other developing countries, and their libraries can be considered as strong information bases.
• Economic and social data: Fundação Getulio Vargas (Rio de Janeiro); Serviço Federal de Processamento de Dados (SERPRO), under the Ministry of Finance; and Fundação Sistema Estadual de Análise de Dados (SEADE), connected with the Secretaria de Economia e Planejamento do Estado de São Paulo.

Baseline data are collected and processed by some of the state environmental agencies. However, if one considers that the generation and processing of environmental data usually depend on sophisticated equipment, it is easy to understand the paucity of monitoring data available to managers, since those agencies are poorly equipped (except CETESB, FEEMA and CEPED). Nevertheless, there are many governmental organizations (not necessarily part of the environmental control structure), which maintain information systems on environment-related subjects. The following represent the main sources of data:

- Secretaria Especial de Meio Ambiente (SEMA), in Brasília, under the Ministério do Interior, which processes selected data collected by several regional and state agencies. It is now responsible for the organization of SIN-IMA, the national system of environmental information, which is expected to be a management information system (if the bureaucracy allows it to go beyond the written plans and acronyms).
- Superintendencia de Recursos Naturais e Meio Ambiente (SUPREN), in Rio de Janeiro, of FIBGE, the national statistical office. SUPREN data banks store data on natural resources, ecosystems, pollution, demography and the economy. As far as national environmental data collection and processing are concerned, SUPREN is actually the best source of information because of its comprehensiveness, methodology and the accuracy of its data.
- Departamento Nacional de Águas e Energia Elétrica (DNAEE) in Brasília, under the Ministry of Mines and Energy, which processes data on both the quantity and the quality of fresh water.
- Departamento Nacional de Produção Mineral (DNPM) in Brasília, under the Ministry of Mines and Energy, which runs two main systems of environmental interest: PROSIG, which uses the Canadian software of the same name to process bibliographic and hard data on Brazilian geology and mining activities and RADAMBRASIL, a special project to collect and to disseminate data on Brazilian geology, geomorphology, pedology, flora and soil use.
- Comissão Nacional de Energia Nuclear (CNEN) in Rio de Janeiro, which maintains a data bank on meteorology, nuclear instruments and other data necessary to implement and run the nuclear power stations now being built in Brazil.
- Instituto Nacional de Meteorologia (INAMET) in Brasília, which runs one of the largest networks for data collection ever established in the country, contributing to its meteorological data bank.
- GEIPOT (Brasília), a permanent working group for research and policymaking on transportation, subordinate to the Ministry of Transports, which collects data on transportation and communication.
- Departamento de Hidrografia e Navegação (DHN) of the Ministry of Navy in Rio de Janeiro, which manages the national oceanographic data bank.
- Laboratório de Computação Científica (LCC) of CNPq (Brasília), a data bank on vegetal resources containing information about over 214,000 species of Brazilian flora.

• Instituto de Pesquisa Espacial (INPE) in São José dos Campos, which is a clearinghouse for satellite images of the Brazilian environment. It processes data from its own research on meteorology, agriculture and natural resources.
• Companhia de Tecnologia de Saneamento Ambiental (CETESB) in São Paulo, which runs a data bank on sanitation and pollution in the State of São Paulo.
• Fundação Estadual de Engenharia e Meio Ambiente (FEEMA) in Rio de Janeiro, which runs a data system on air and water quality.

Brazilian information users are now connected on line with foreign data bases through (Serviço Internacional de Communição de Dados - Brazilian International Service of Data Communication) INTER-DATA, the national network linked to Timeshare Network (TYM-NET). However, the reverse connection is not feasible yet.

Even though the Brazilian panorama of environmental information looks well organized, it is actually very incipient for a series of reasons, such as the low priority of environmental matters within governmental policy, lack of prestige of SEMA, paucity of qualified human resources, insufficient financial support and internal disputes among environmental agencies.

Mexico

In Mexico, as in Brazil, science policy is set by a national council, which is a part of central government. CONACYT, the Mexican council, is also entitled to formulate scientific and technological information policy, including its policy on the environment. Thus, CONACYT should be considered the first referral source for documentation centers, scientific and technological organizations and for experts on the environmental sciences. After CONACYT, the UNEP Regional Office, Secretaria de Salubridad Asistencia, Secretaria de Asentamientos Humanosy Obras Publicas (SAHOP), Centro de Ecodesarrollo and PAHO's Centro Panamericano de Ecología Humana y Salud (ECO) can be considered useful referral sources for more specific questions.

As to documentation, the main institutional sources are as follows:

• CONACYT, which has organized the national union catalog of periodicals, the computerized *Directorio de Fuentes y Recursos de Información Documental* (DIFRID) and the national bibliography on marine sciences. Subor-

dinate to CONACYT are two documentation centers on ecological subjects: Instituto de Ecología and Instituto de Recursos Bióticos.
- INFOTEC, the Center of Information and Documentation for Industry, co-ordinates technological information systems throughout the country, besides directly performing some services of information consultancy and technology administration. Those services are based on two chief sources: its library and the foreign data bases accessed on line by INFOTEC. Its library is quite strong on industrial pollution (including both Mexican and foreign literature) and is planning to take the responsibility of publishing the national bibliography on industrial development and technology—*Bibliografia Industrial de Mexico*—published through 1976 by Banco de Mexico.
- Instituto Nacional de Investigaciones Nucleares, for national literature on nuclear energy.

Like Brazil, Mexico is also connected with TYMNET for on-line access to foreign data bases. CONACYT indirectly controls the access through its intermediate service, the Servicio de Comunicación a Bancos de Información (SECOBI), an information bureau which offers better prices than users would pay for an independent search.

With regard to hard data about the state of the Mexican environment, the following organizations collect and process data on their specific area of interest:

- Secretaria de Salubridad y Asistencia, mainly through the Subsecretaria de Mejoramiento del Ambiente, which has information on food contamination and noise, air and water pollution. Its strongest area seems to be air pollution control, through IMEXCA, the Mexican index of air quality, which is monitored by a large network.
- Secretaria de Agricultura y Recursos Hidraulicos, mainly through its Dirección General de Protección y Ordenación Ecológica, which maintains two information systems, one on water quality (SICA) and the other on the uses of water (SIUA).
- SAHOP, which deals with data on sanitation, water pollution, urban and rural development, ecological surveys and human settlements.
- Instituto Mexicano del Petroleo, which stores specific data on the production, commercialization and consumption of petrol and on pollution caused by activities related to petrol exploitation.
- Secretaria de Programación y Presupuesto, mainly through its statistical office, which holds economic and social data.

A still unclear institutional panorama and weak coordination are main obstacles in finding out about environmental information in Mexico.

AFRICA

UNEP, the UN Economic Commission for Africa (ECA) and the Organization of African Unity (OAU) may be considered the major referral centers for sources on the African environment and the most important producers and organizers of substantive information on the subject.

ECA, for instance, has established in Addis Ababa, Ethiopia, the African Regional Center for Technology in collaboration with OAU, UNIDO, UNCTAD, World Intellectual Property Organization (WIPO), ILO, UNESCO and FAO. OAU maintains its Scientific and Technological Research Center in Lagos, Nigeria. Both centers' ideals lean toward appropriate technology within the ideology of ecodevelopment. ECA's Pan-African Documentation and Information System (PADIS) has implemented a data bank of information and statistics on the African environment. Information on the African social environment may be obtained from the Center for the Coordination of Social Sciences Research and Documentation in Africa South of the Sahara (CERDAS) in Zaire, which has published *A Directory of Scientific Research Institutions in Zaire* (the enterprise is to be extended to the institutions of the whole region), and *A Directory of Archives, Libraries and Documentation Centres in Zaire*. However, the best sources of information on African NGOs will still be provided by the Environmental Liaison Center (ELC) in Nairobi. To complement the picture, UN Cairo Demographic Center in Egypt may help in the aspects concerning African population data.

Five more groups of organizations interested in the African environment can be identified that have a national scope. Their areas of environmental action are:

- Appropriate technology: the Appropriate Technology Unit of the Institute of Agricultural Research in Ethiopia and the Appropriate Technology Information Center of the Department of Agricultural Engineering at the University of Ife in Nigeria.
- Development and ecodevelopment: the National Institute of Development and Cultural Research in Botswana, Documentation and Communication Service for Development in Ethiopia, Societé Africaine d' Études et de Développement in Upper Volta and Technology Relay for Ecodevelopment and Planning in African Environments in Senegal.
- Human settlements: African Rural Housing Association in Ethiopia and Cen-

ter for Research and Development in Housing, Planning and Building of the
Faculty of Architecture at the University of Science and Technology of Ku-
masi, Ghana.
• Environmental health and pollution: High Institute of Public Health of the
Alexandria University in Egypt and National Research Center in Egypt.
• Agriculture and rural development: African Rural Storage Center in Nigeria,
East African Agricultural and Forestry Research Organization in Kenya, In-
ter-African Bureau for Soils of the OAU in the Central African Republic and
Office National de la Promotion Rural in the Ivory Coast.

Egypt

Although Egypt has an old tradition with science, finding local scientific and technological information in the country is a hard job for a
foreign researcher because of the language barrier, secrecy blockade,
lack of institutional coordination or leadership and consequent dispersion of sources. In spite of these hindrances, some local organizations
have a potential for working as information sources if the enquirer meets
the right person; the serendipity factor is crucial for success in research.

Science policy is formulated by the Academy of Scientific Research
and Technology, which also delineates information strategies to fulfill
those policies. As far as ecological matters are concerned, the National
Council for Environmental Protection is the Academy's body responsible for major decisions, in close connection with the High Committee for the Protection of the Environment. Thus, the Academy of Scientific Research and Technology is considered a natural referral source
of environmental information in Egypt. Furthermore, this referral
function is emphasized by the fact that the Environmental Information
Office of the National Council for Environmental Protection is an IN-
FOTERRA focal point in the country. Unfortunately, information officers understand that they must act as protectors of sources against
enquirers, instead of acting as real intermediaries. Such an attitude im-
poses serious drawbacks to research.

Two helpful referral sources on Egyptian environment are the National Research Center, because it embraces many areas of environmental interest, and the High Institute of Public Health of the Alexandria University, whose personnel are well informed about the whole
institutional panorama of the Egyptian environment.

A directory of Egyptian scientific organizations and experts already exists, prepared by the statistical department of the Academy of Scientific Research and Technology. The National Information and Documentation Center (NIDOC) now prepares the *Directory of Scientific and Technical Libraries in the ARE*, and published in 1980 the *Inventory—Ongoing Research Projects*.

Data are being generated through research and monitoring by many organizations, some of them inclined to share information with outside users. Regardless of openness and goodwill, the following organizations store some kind of environmental information:

- National Research Center provides information mainly through its specialized bodies, such as the Air Pollution Laboratory, Water Pollution Control Laboratory, Ecological Research Division, Office for Technology Transfer and Solar Energy Laboratory.
- General Organization for Sewerage and Disposal Control collects data on sanitation and water pollution of both domestic and industrial wastes.
- General Organization for Industrialization (GOFI) holds information mainly on industrial development, industrial pollution and marine pollution.
- Egyptian General Organization for Geological Survey and Mining Projects aims at land and mineral resources.
- Ministry of Petroleum has information on marine pollution by petrol, especially the studies related to the MEDLINE project, undertaken in conjunction with Arab Countries, to clean the Mediterranean and the Red Sea. Egyptian General Petroleum Company, the Petroleum Research Institute of the Academy of Scientific Research and Technology and the Ministry of Maritime Transport's National Committee for the Combat of Marine Pollution hold complementary information on the subject.
- General Organization for Building, Housing and Planning Research produces information on housing, reconstruction, community development and alternative building material.
- Ministry of Agriculture, mainly through its Soils and Water Research Institute, offers environmental information related to agriculture.
- Desert Institute has information on Egypt's research and actions in favor of dry soil recovery.
- Center for the Development of Engineering and Industrial Design performs research and training in industrial technology.
- Atomic Research Center acts in the area of nuclear energy information.
- Ministry of Health holds information on public health and builds up a data bank out of information collected by a national monitoring network of air quality.
- Remote Sensing Center, affiliated with the Academy of Scientific Research

and Technology, is the clearinghouse of Landsat images of Egyptian territory.

- Central Agency for Public Mobilization and Statistics (CAPMAS), the main statistical office of central government, holds information on population, economy, housing, education, employment, cultural statistics (production of books, number of libraries and so on) and on many other socioeconomic aspects of the Egyptian environment. A very important information source, it is deplorable that the director of CAPMAS closes the organization to foreign researchers.
- Academy of Scientific Research and Technology, mainly through its Nile River Data Bank. Built up to give information support to the Lake Nasser Project, this data bank monitors the Nile River's quality of water, too.
- Al-Azhar University, mainly through its Islamic Center for Population Studies and Research.

Since data sources are many and uncoordinated, the National Committee for the Protection of Environment plans to build up, by computer, a general environmental index of available information in all Ministries, a kind of union list of environmental data. In recognition of the inadequateness and inaccuracy of existing data, some organizations—namely, GOFI, CAPMAS and the General Authority for Arab and Foreign Investment—plan to build up their own management information system (MIS).

Concerning Egyptian bibliography on the environment, most of it is published in Arabic. The main sources of information are:

- National Information and Documentation Center (NIDOC), which maintains the best scientific and technological library of Egypt (including the National Reference Library of Science and Technology). It has organized the Sinai Data Base of bibliographic information about social, political and physical aspects of that region. NIDOC also publishes many scientific periodicals, the abstracts of conferences held at the Academy and the national scientific bibliography. It has also been publishing the section "Arab Science Abstracts" within the *NIDOC Documentation Bulletin*, covering scientific literature produced in the Arab world. Another useful source of information will be the *Union Catalogue of Scientific and Technical Serials in the ARE*, now under preparation by NIDOC.
- Cairo University, which hosts part of the National Reference Library of Science and Technology.
- National Research Center's Library mainly holds documents related to research undertaken at the Center and at its specific laboratories.

ASIA AND THE PACIFIC

Considering the environment of developing countries situated in Asia and in the Pacific, after the INFOTERRA directory, one can still be helped by the UNEP Regional Office in Bangkok, Thailand. Then, sources can be found through both the UN Economic Commission for Western Asia (ECWA) in Lebanon and the UN Economic and Social Commission for Asia and the Pacific (ESCAP) in Thailand. As part of their information work, ECWA has established the Center for the Transfer and Development of Technology, and ESCAP has created the Regional Center for Technology Transfer, both of them searching for environmentally sound alternatives of development.

Various national organizations, starting with the central agency of environmental management of each country, also may be useful information sources. The following ones are a sample:

- Agriculture and rural development: Bangladesh Agricultural Research Council, Center for the Development of Human Resources in Rural Areas in the Philippines and the Office of Village Development of the Department of the Prime Minister in Papua, New Guinea.
- Ecodevelopment and appropriate technology: Center for Science and Environment in India, Ecodevelopment Cluster in Iran, Appropriate Agricultural Technology Cell of Bangladesh Agricultural Research Council, Indian Department of Science and Technology (mainly its Council for Scientific and Industrial Research), South Pacific Appropriate Technology Foundation in Papua, New Guinea, and the Appropriate Technology Information Service of Indian Institute of Science.
- Scientific and technological development: Bangladesh Council for Scientific and Industrial Research, Development Technology Center of the Institute of Technology Bandung in Indonesia, Indian Institute of Science, and Technology Transfer Center of the Korean Institute of Science and Technology.
- Environmental and public health engineering: National Environmental Engineering Research Institute (NEERI) in India, Institute of Public Health Engineering and Research of the University of Engineering and Technology in Pakistan and School of Environmental Sciences at the Jawarharlal Nehru University (JNU) in India.

India

Since Indian scientific and technological development is the highest among developing countries, it is not surprising to find in that country a well-organized structure of levels of R&D, technology transfer and implementation, resulting in a vast number of information sources that can help a researcher interested in the Indian environment.

Science policy is formulated by the Science Advisory Committee for the Cabinet (SACC), under the Prime Minister, in agreement with the Department of Science and Technology (DST). The Department of Environment (DOE) is responsible for implementing the environmental policy of the government; such a policy should be considered as a new ecological dimension to orient all programs and projects in every Ministry and Department.

The policy of scientific and technological information is the responsibility of DST, as the national focal point and coordinating body of the National Information System for Science and Technology (NIS-SAT), which includes environmental information as well.

Linked to DST is the main complex of laboratories and research centers, most of them under the Council of Scientific and Industrial Research (CSIR) in New Delhi, which should be considered a comprehensive referral channel of environment-related sources. Its National Environmental Engineering Research Institute (NEERI) in Nagpur is prepared to render specialized information. However, the major referral source of environmental information in India is DOE in New Delhi, which works very efficiently as the INFOTERRA national focal point in the country. Three more organizations could complement the array of information institutions: the School of Environmental Sciences at JNU in New Delhi, the Center for Science and Environment in New Delhi, and the National Research Development Corporation of India in New Delhi.

There are also available many sources on science and technology organizations and experts. The main ones are: *Directory of Scientific Research Institutions in India* (published by the Indian National Scientific Documentation Center [INSDOC] in 1969), National Register of Scientific and Technical Personnel (periodically organized by CSIR since 1948), Indians Abroad Register (organized by CSIR since 1958), Specialists Register (organized by CSIR since 1975), *Current Research Projects in CSIR Laboratories* (published by INSDOC, 1972 and 1976), *Directory of Indian Scientific Research in Indian Universities* (published by INSDOC, 1974) and *Directory of Testing Facilities in India* (published by INSDOC, 1981). The above sources have been constantly updated by responsible organizations. New data are available at their facilities, even when those data have not yet been published.

Collected from independent surveys, data on existing information services in India are available at INSDOC in New Delhi, at the Documentation Center of the Tata Energy Research Institute in Bombay and at the Documentation Research and Training Center (DRTC) of the Indian Statistical Institute in Bangalore. Besides these already accessible sources, some institutional data files are under preparation: ongoing research in science and technology, R&D facilities and testing facilities. DOE has also published a small directory, *Projects on Environmental Research*, with 1981 data.

As far as data on the Indian environment are concerned, central government in its *Sixth Five Year Plan, 1980–85* has proposed the establishment of an Environmental Information System (ENVIS), one of the above mentioned MIS, made up of data collected by existing organizations. For the time being, two initial sources may be consulted: the *Wealth of India*, an encyclopedia of Indian raw materials and industrial products organized and published by CSIR's Publications and Information Directorate (PID) and the *Directory of Data Banks*, published by INSDOC and DRTC in 1977, which identified 181 data files (manual and computerized) in the country by that time. Concerning institutional sources, DOE and CSIR can be consulted successfully. A third organization, with great potential to inform on the management information systems now being implemented in some governmental bodies, is the National Informatics Center of the Electronics Commission in New Delhi. It has been commissioned by central government with the responsibility of developing and implementing such systems. Some of those data banks are being set up under NISSAT, including one on agriculture and environmental information and another on industry and technology information. Environmental data from DOE and the Department of Forestry and bibliographic data from INSDOC are already being processed at the Center. Although the National Informatics Center is the government body responsible for the development and transfer of informatics technology (software and hardware) for data processing and communication in India, and considering that the Data Banks Development project is funded by UNDP, the Center does not render information to foreign researchers. The same curtain of secrecy is interposed by the director of the Center in relation to India's on line facilities for international data exchange through satellite, although information on INSAT, the Indian National Satellite System for Tele-communication, Television and Meteorology, and on APPLE, the Ariane Passenger Payload Experiment, have already been made public through other sources, such as members of the Indian scientific community, technical magazines, newspapers and even in the pages of *Swagat*, the magazine of Indian Airlines on international flights. The most recent information that the above mentioned bureaucrat could offer, on specific request, was the booklet *Electronics Information Potential in India*, published in 1968. Equally secretive seems to be the Indian Ministry of Health in relation to foreign researchers, having also refused to contribute to this present investigation. This picture may suggest to foreign researchers that information concerning the National Informatics Center's and the Ministry of Health's areas of action should be looked for through indirect sources where national security is possibly not considered at such a superficial level.

Last, there are still organizations that deal with specific data. These main environment-related institutions are grouped here under five areas:

- Social environment (population, health, family welfare, human settlements): Indian Statistical Institute in Calcutta, Indian Council of Social Science Research in New Delhi, National Institute of Health and Family Welfare in New Delhi, Indian Council of Medical Research in New Delhi, CSIR's Central Drug Research Institute in Lucknow and Ministry of Works and Housing in New Delhi.
- Natural environment: Indian Meteorological Department; National Remote Sensing Agency of India; India Space Research Organization in Ahmedabad; Geological, Botanical and Zoological Surveys; CSIR's National Geophysical Research Institute in Hyderabad; CSIR's National Botanical Research Institute in Lucknow; Department of Forests; Indian Agricultural Research Institute; CSIR's National Institute of Oceanography in Goa; Central Soil and Water Conservation Research and Training Institute; CSIR's National Committee for UNESCO's International Hydrological Program; Central Ground Water Board; and Kerala Sastra Sahitya Parishad (KSSP) in Trivandrum.
- Environmental pollution: Department of the Environment in New Delhi, CSIR's Industrial Toxicology Research Center in Lucknow, CSIR's National Environmental Engineering Research Institute (NEERI) in Nagpur, CSIR's National Institute of Oceanography (branch of Bombay), Central Public Health and Environmental Engineering Organization of the Ministry of Works and Housing in New Delhi, Central Board for the Prevention and Control of Water Pollution in New Delhi, Central Board of Insecticides and Pesticides in New

Delhi, School of Environmental Sciences at JNU in New Delhi and Banaras Hindu University.
- Energy: Tata Energy Center in Bombay, CSIR's Indian Institute of Petroleum in Dehra Dun, Ministry of Petroleum in New Delhi, Bhabha Atomic Research Center in Bombay and the Department of Science and Technology in New Delhi.
- Technological development and AT: India National Science Academy in New Delhi; Department of Science and Technology in New Delhi; Indian Institute of Science in Bangalore; National Research Development Corporation of India (NRDC) in New Delhi; Indian Institutes of Technology in Bombay, New Delhi, Kanpur, Madras, and Kharagpur; Directorate-General of Technical Development in New Delhi; CSIR's Central Food Technological Research Institute in Mysore; CSIR's National Chemical Laboratory in Pune; Appropriate Technology Cell of the Ministry of Industry and Civil Supplies in New Delhi; Center for Science and Environment in New Delhi and KSSP in Trivandrum.

To find out about national bibliography on the Indian environment, one should look through more generic sources on science and technology or directly consult the libraries and publication facilities of the above mentioned organizations. Within this informational frame, the following institutions necessarily should be looked for: NISSAT headquarters at DST in New Delhi, INSDOC in New Delhi, the National Science Library in New Delhi, the National Library in Calcutta, the Indian Agricultural Research Institute Library in New Delhi and JNU libraries in New Delhi.

One may infer that most of the national bibliography on the Indian environment (and on environmental sciences) is available in New Delhi, a condition that facilitates research and transforms this city into a privileged place for Third World environmentalists to perform further studies on the subject. Some bibliographic tools are also available, such as the *Union Catalogue* (under preparation at INSDOC and already includes eighteen main libraries of the country; computer lists are always available), *Directory of Indian Scientific Periodicals* (published by INSDOC, 1968 and 1976), *Indian Science Abstracts*, and the broader *Guide to Current Literature in Environmental Engineering and Science* (published by NEERI, monthly), which includes both Indian and international papers.

A final observation should be made about the concern of Indian government for the transfer of appropriate technology to other developing countries. This is especially noticed at the National Research Development Corporation of India, which has established its Center for Technology Transfer in order to assist managers in India and other Third World countries.

POLICY AND POLITICS

Environmental policy of a country can first of all be inferred from some indicators presented in previous chapters, such as the kind of legislation enacted to manage the national environment, the level of freedom experienced by both media and the public in discussing and demanding higher quality of life for the whole population, the degree of public participation welcomed by government and the availability of local information to every citizen interested in improving the common habitat. In discussing these aspects, the environmental ideology was also implicit in the text. Nevertheless, government orientation and interinstitutional competitiveness among Brazilian environmental organizations will be hereafter discussed and compared with the scene in Mexico, India and Egypt to illustrate the matter. Explicit environmental policy and politics in those countries will either confirm or deny earlier assumptions with reference to common patterns of underdevelopment and to the correlation of environmental ideology with the adopted style of development.

Brazil

From the fifties on, the central government of Brazil has adopted, with the support of foreign capital and technology, a developmental policy where economic growth and industrial advancement prevailed over social enhancement. By electing such a style of transnational dependent capitalism, the associated environmental problems (consumerism, loss of cultural authenticity, overexploitation of natural resources, industrial pollution, conurbations and so on) have unavoidably been adopted. These new problems, combined with existing malaises of underdevelopment, led the government to find room for people's move toward a better living condition in a consequently improved environment.

The definition of an internal policy for the management of environment has been a complex job for Brazilian authorities, walled between their own narrow developmental policy and glimpses of the actual situation, which demand a pluridimensional attitude constituted by social, conservationist and ecodevelopmental aspects of concern. Such a complexity can be seen through the consideration of some realities of the country:

- A vast proportion of the population still lives under the subsistence level, struggling against the pollution of poverty, that is, malnutrition, epidemic diseases, lack of sanitation and lack of basic education. This aspect should be a social concern of policymakers.
- An enormous part of the country is not yet sufficiently inhabited or exploited economically in a way that would benefit the majority of the Brazilian population. This topic demands both conservationist and social concerns from the policymakers.
- The country's dedication to economic development has led some areas to industrial pollution and lowered the quality of life of urban centers. Policymakers should work in close cooperation with technologists and economists: The country's lack of condition to select the best technology for local conditions has undermined both the social and natural environment. An ecodevelopmental strategy should be sought by policymakers.

Before the seventies, the Brazilian government had no specific policy for pollution control. The subject was vaguely approached as a part of sanitation and public health. The first governmental document to reflect publicly an explicit policy was Law 5.318 of July 26, 1967, which contained the main directives intended for governmental actions and the main goals, as a statement of the national sanitation policy. It embraced water supply, sewage works, solid waste disposal, environmental pollution, flooding and erosion. Since industrial pollution is a problem of the seventies, the first time it came about in governmental plans was in 1970 in the *Metas e Bases para Ação do Governo*.[1] This official document contained the goals and guidelines for federal government in the period 1970–71 and stressed health and sanitation, education and technological development (mainly industrial). As part of health plans, it included the need to control environmental pollution in urban areas, mainly São Paulo and Rio de Janeiro.

The next governmental plan was the *I National Plan of Development* (I PND), for the period 1972–74.[2] It emphasized economic and technological development. To implement this plan, special programs were formulated in the *I Basic Plan of Scientific and Technological Development* (I PBDCT).[3] Under these plans, the prevention and abatement of pollution and the overall environmental concern are only marginal topics, through sanitation, public health and agricultural technology. This is understandable when one remembers that the Minister of Planning of Brazil at that time declared unofficially to American news agencies in 1972 that "Brazil could happily become an importer of pollution, as it has a lot left to pollute." Officially, through the Ministry of Foreign Affairs, central government denied the publication and declared that:

- Brazil supports international movements in favor of the environment, like the Stockholm Conference, but it believes that developed countries are responsible for world pollution and its abatement.
- There is a need for controlling industrial pollution in some metropolitan regions in Brazil for which special policy is necessary. But the Brazilian environment has a large carrying capacity for pollution because of its huge unoccupied areas.
- Economic growth and industrial development should have priority over protection of the environment in a developing country.[4]

Through the years, this has been Brazil's overall policy. Although very technocratic, one positive feature of this government was the creation of SEMA in 1974 in response to the Stockholm Conference.

The period 1975–79 has been guided by the *II National Plan of Development* (II PND) and its corresponding *II Basic Plan for Scientific and Technological Development* (II PBDCT). Both of these focus on the preservation of the environment, control of pollution and urban development. In opposition to previous governments, these plans state an environmental policy, the main line of which is a compromise between economic development and the protection of environment, that is, the national use of natural resources, promotion of social welfare and industrial development, avoiding pollution and deployment of the environment.[5, 6] The basic legislation for the control of industrial pollution was adopted during this government.

From March 1979 on, Brazil has had another government, whose plans, III PND[7] and III PBDCT,[8] have emphasized the quality of life to be reached through a balanced socioeconomic development of all regions, the smoothing of social inequalities, the exercise of democratic freedom by society and the conservation of natural resources.

Contrasted with the explicit statements of formal policy, some points may highlight the real environmental policy implicit in this government's current actions along its mandate:

- Even though the military is still in power, the President has stated his intention of gradual liberalization and redemocratization of the country. This has been happening slowly, in spite of some right-wing reactionary forces, which eventually call back nondemocratic measures, such as the censorship imposed in the area of Brasília in 1983 regarding meetings which would manifest public reaction against the government economic policy.
- The President had declared early in his government that development plans for the Amazon region would be guided by the policy of preserving its ecological equilibrium,[9] but its exploitation goes on savagely.[10], [11]
- The present economic policy adopted by the government is in favor of severe economic recession and nongrowth, as commanded by the International Monetary Fund (IMF), the new external "ruler" of the country. Such a recessive policy has been generating worse environmental imbalance via unemployment, misery, hunger, disease, violence and all kinds of social maladjustments.
- The national budget of 1981 reserved only .08 percent to pollution control (the only environmental sub-area granted support), and for 1982 the whole area of environment (including natural resources) received no more than 5.8 percent.

The Secretary of Environment, when interviewed for the present research, stated SEMA's belief and political orientation:

- Management of the environment by the Brazilian government must include the improvement of socioeconomic conditions of the population, the surveillance of natural resources and the control of pollution. It means that government should look after the problems derived from underdevelopment and from development.
- Environmental control is a decentralized function: each state is responsible for local control, and SEMA's responsibility is to help or intervene whenever necessary.
- The exchange of environmental information and technology among countries should be a common pattern for governments, and Brazil, as a developing country, should not be afraid of having its interests and aspirations threatened by developed countries under the guise of environmental concern.

An official document prepared by SEMA is more modest than the previously mentioned III PND and hence closer to the true situation, where formal policy meets reality. According to this document, SEMA concentrates environmental policy on preserving the quality of life in urban centers (mainly in metropolitan areas), by dispersing industries and by supplying cities with infrastructure for a safe water supply, sanitation and solid waste disposal. Even more clearly, SEMA specifies that the Brazilian environmental policy is founded on the power of governmental legislation.[12]

A look into the institutional politics will reveal that SEMA has plenty of goodwill and plans, but it lacks both power and resources (budget and qualified personnel) and for these reasons does not achieve the expected results. The following examples illustrate this handicap:

- SEMA's sphere of action has been overlapped by other organizations at the federal level, for instance, by SUPREN.
- Official information is often denied to SEMA, as is the case in its relationship to the Departamento das Nações Unidas (DNU) of the Ministry Foreign Affairs.
- Links between SEMA and environmental state agencies are quite informal and weak, a situation aggravated when one of these agencies (like CETESB, in São Paulo) chooses to maintain an independent scheme and can do so because it is richer and possesses better technological knowledge and equipment.

Concerning Brazilian foreign policy related to environment, some basic lines can be identified at present:

- The international concern and actions of any supranational organization should not trespass upon national sovereignties but must accept the national development frameworks as a fundamental parameter.
- Brazilian policy has evolved from opposition to the international environmental movement to a compromise, that is, development in a clean environment. In order to achieve this, the country is interested in the development of appropriate technologies.
- Brazil neither supports the "doomsday" ideas of the Club of Rome nor the belief in the inexhaustibility of world resources.

The comparative analysis of central government policy and the specific orientation followed by environmental organizations in the government (mainly SEMA) shows that a subjacent conflict between technocracy and quality of life still prevails among governmental institutions.

The above policy and politics reflect on the environmental information scene: among such dispersion and dissonance, SEMA's efforts have still not succeeded in coordinating national resources into an actual network of environmental information. Overlapping areas of action and resources waste are quite frequent, caused mainly by the absence of a clear policy or environmental information.

In a broad context, Brazilian authorities included an information policy for the first time in the official government program[2] in 1972, to be implemented through the specific projects drawn by the I PBDCT.[3] As far as environmental information is concerned, governmental plans for the period 1975–79[6] have included some specific projects in the areas of natural resources, agriculture, health, fishery resources, energy and industrial technology.

Environmental information is not included in the present government's plans.[7], [8] Only indirectly does the III PND mention the government priority as to dissemination of information on the protection of environment and natural resources. Recently implemented in June 1983, Law 6.938—the main environmental legislation at federal level—includes the diffusion of environmental technology, data and information as one of the objectives of the national environmental policy. The same document creates the National Information System on the Environment, which will constitute the basis for the implementation of a national policy.[13]

At the international level, it seems that Brazilian policy concerning environmental information, determined by the Ministry of Foreign Affairs, has initially been in favor of getting the maximum from the world information potential and contributing the minimum. The alleged reason was the government fear of foreign interference in the internal management of the country's environment, which is quite a paradox for a government that surrenders national sovereignty to external financial institutions in the international market. The government has evolved from this illiberal position, however, and today Brazil participates in two of UNEP's subsystems (INFOTERRA and IRPTC), in the REPIDISCA, in the DOCPAL, in the INFOPLAN and in the AGRINTER. But it still resists UNEP's GEMS for reasons of national security and self-determination, allegations made since the creation of this subsystem in 1973.

Notwithstanding its marginal inclusion in developmental programs of the government, environmental information in Brazil is still in its infancy, and its development will probably depend on some extrinsic factors, which was the object of discussion in chapter 3 about the media and public power, in chapter 4 about the design of the environmental system and in the present chapter on the institutional organization. These factors can be summarized in four main points: (a) the political and economic destiny of the country in the coming years, (b) the policy adopted by government in the management of the environment, (c) internal politics amongst Brazilian environmental agencies and (d) the development of information and communication technology in the country.

Thus, it is thought that any definition will only come as part of an overall information policy of the country within a new democratic state.

Mexico

Mexico has had a very active political line concerning environment. Nevertheless, as in the case of Brazil, the Mexican government has an explicit policy that especially favors the environment, while a technocratic bias is clearly implicit in different documents that tend toward economic development.

In the official report to UNEP, Mexico's government declares part of its policy is to have as its goal the improvement of living conditions of the population. The guiding points of such a policy are said to be a regional approach to environmental problems, governmental responsibility in preserving the environment from degradation, ecodevelopment, public participation and international exchange.[14] This is certainly true, but another document, which reviews Mexican government actions in the management of its environment along the last ten years and the implicit policies adopted, differs somehow, giving the impression that the economic policy determines the environmental one. Some of the observations the document contains are that (a) the betterment of environmental quality should be adequate to each country's previous determination toward development and industrialization, (b) economic policy is the main determinant of environmental sanitation activities and (c) the satisfaction of governmental economic policy is a guiding principle of environmental education.[15] These observations seem to contradict the Mexican environmental position reported to UNEP.

On the other hand, by observing Mexican legislation, one will find another indicator of the internal environmental policy actually adopted by the country. This indicator matches the formal policy by showing that government not only punishes industrial pollution but also gives incentives to factories in order to prevent their contamination of the environment. The government and private sectors are in this way integrated, although the intervention of the public sector to conserve natural resources (for example, the early nationalization of petrol) predominates.

The strategical programs of the central government to implement its environmental policy and the actual plans and projects being executed by various agencies, however, reveal the seriousness of the government in reconciling socioeconomic development and environmental quality. Among these, the ecoplans for human settlements deserve special mention.

Institutional politics seem to work well enough because Mexican legislation allows for the creation of intersecretarial commissions to study specific problems. Thanks to that, interested bodies from various Secretariats come together officially, regardless of the power that the inviting organization possesses (for example, Subsecretaria de Mejoramiento del Medio Ambiente, for environmental matters).

As far as international policy is concerned, Mexico defends that the solution of environmental problems should be looked for through international cooperation. Thus, it collaborates actively with international organizations besides taking part in international programs, such as the Environmental Project for the Greater Caribbean.

Within the domain of the Secretariat of Foreign Affairs, the Stockholm Declaration is the main guiding principle, followed by the defense of national sovereignty with regard to the country's environment. The Secretariat requests cooperation from Mexican scientists and university lecturers in order to discuss technical matters, but political themes remain its only responsibility. No control at all is imposed against communication of Mexican environmental organizations with their foreign partners, as it happens sometimes in Brazil. The relationship between the Secretariat of Foreign Affairs and Mexican environmental organizations is undoubtedly excellent, and governmental policy is certainly behind it. This characteristic entitles Mexican embassies to be especially recommended as diligent information agents on the Mexican environment.

The Mexican scientific information policy, established by CONACYT, aims at supporting development through scientific and technological information. In this policy, not even marginally has environmental concern been taken into account. The Subsecretaria de Mejoramiento del Ambiente has not formulated a policy on such a matter either. However, actions of the various organizations in the environmental structure show they care about information transfer. The special case of SAHOP links governmental policy of ecodevelopmental content to the information scene by having a strong program on information transfer of appropriate technology to the general public as part of its ecoplans for human settlements and self-sufficient homes.

At the international level, Mexico participates in UNEP's INFO-TERRA, GEMS and IRPT, besides taking part in the REPIDISCA, LATINAH, INFOPLAN, BIREME and AGRINTER.

Considering the difficult situation of the Mexican economy and the country's internal social problems, which demand great investments and political attention from government, the near future of environmental information there is not clear. But these present observations reveal an ecodevelopmental tendency which may come to be explicitly adopted by the Mexican government.

Egypt

From both political and physical viewpoints, Egypt is in a very vulnerable condition; this makes the role of policymakers especially delicate. Under such constraints, planners who designed *The Five-Year Plan: 1978–1982*[16] were wise enough to take as its ideological basis the restoration of a genuine balance among the various segments of Egyptian society.

This overall policy gave priority to military preparedness in defense of Egyptian land and of the Palestinian cause. Following this priority, the planners aimed at increasing more rapidly the rates of economic development, which they considered ''a matter of life or death for Egypt.'' Complementarily, the ''open door'' policy was introduced to give Egyptians access to modern technology in order to speed up industrialization. Both technology and capital were then welcomed from abroad at the same time that the Egyptian private sector was also invited to share ''the burdens of development.''

Environmental policy was not explicit in *The Five-Year Plan: 1978–1982*, but benefits to this sector would come as a result of other socioeconomic priorities, such as agriculture, irrigation and drainage, housing, drinking water, sewage and health services. This attitude of considering environmental quality only as a consequence of other measures can be confirmed through the sectoral policy of the Ministry of Industry and Mineral Wealth, which considers industrial development as the key to improving the standard of living.[17] Industrial pollution control is approached by this authority through land utilization control, better technology and recycling of wastes.

Concerning information, there is not a real policy but only isolated efforts to process and to disseminate mainly technological information, even though industrial organizations complain about the lack of data and documents.[18] On the other hand, it seems that scientists and technologists lead the scene of information transfer. Thus, for its institutional position, the Academy of Scientific Research and Technology has given support to scientific and technological information, and any formulation of environmental information policy will naturally come from the Academy.

The analysis of Egypt's policies—global, environmental and for information—denudes a desperate effort of the country toward development. For the attainment of this aim, priority is given only to measures that directly contribute to a faster economic growth. So it seems that the environment and information transfer are still considered marginal to developmental plans within the planners' conceptions.

India

India's government policy, as in the previously analyzed countries, aims at development, but within the frame of a socioeconomic equality[19] and total human betterment.[20] On the other hand, it is transparent even to a foreign observer that scientific and technological development in India is neither a matter of chance nor the accumulated result of uncoordinated projects. It is actually the result of a clear policy, explicit in the quinquennial plans of the central government. In both the *Fifth Five-Year Plan* and the *Sixth Five-Year Plan, 1980–85*, science and technology are seen as a means to achieve self-reliance and a better quality of life for the local population, mainly for those people from backward areas.

Equally clear is the information policy, considered by the government as "a major component of policy for science and technology" to be "used in guiding social evolution."[21] In the *Sixth Five-Year Plan*,

1980–85 information policy regarding MIS was also formulated to provide a scientific basis for planning in addition to specific policy for scientific and technological information.

Environment comes under the central government policy via two different approaches:

- The environmental sciences are dealt with as part of overall scientific and technological policy in both plans. In the *Fifth Five-Year Plan* they appear under each sector (for example, natural resources and marine resources), and the development of scientific knowledge through research and education is specifically considered.
- The management of the environment is seen in the *Sixth Five-Year Plan, 1980–85* not as a specific area but as a new ecological dimension of the country's management toward development, to be adopted by all sectors of the government. Furthermore, concern for environment is considered as a precondition for sustainable development.[22]

One could wonder whether the adoption of such an ecodevelopmental pattern of the country's management is not associated with the fidelity to a style consonant with the local culture. Such a philosophy of the government has been stated by the Indian authorities on different occasions, as for example by Indira Gandhi during the Stockholm Conference, when she said that "a higher standard of living must be achieved without alienating the people from their heritage."[23]

Within the holistic concept of environmental management elected by the government, the environmental information policy is intended to support this environmentally sound development and to improve the living conditions of the whole population. The result of such a policy is environmental information being disseminated to various audiences (all sectors of government, private production groups and the lay population) under different nuances of content and levels of complexity. Thus, the *Sixth Five-Year Plan, 1980–85* has an explicit policy for environmental information, besides having included dissemination of environmental sciences under the policy of scientific and technological information.

Another facet of India's policy is that, by adopting a culturally consistent pattern of development, environmental research and information groups are deeply involved with the development and transfer of alternative technologies that are regionally appropriate. Probably because of the very power of this holistic policy, governmental bodies interrelate peacefully and efficiently, and DOE's coordination seems to be well accepted. Both the policy for environmental management and the environmental information policy include public participation, which has been motivated through direct action of DOE and intense participation of NGOs. Concerning public participation among the analyzed countries, India is the only one to include in its information policy the objective of creating information consciousness among users.[24]

Although not consolidated yet, it is already clear that the ecodevelopmental ideology, with an intrinsic social component, strongly underlies the policy adopted by the Indian government. Very positive results for both the population and the environment will certainly come soon.

Comparing the Approaches

As suggested in chapter 4, those indicative lines of environmental information policy—if not made explicit by government—may be identified from the overall policy adopted by a given country in the management of the natural and social environment. Thus, countries like Zaire and Tanzania are likely to adopt a conservationist policy concerning environmental management and information, the same way Nigeria and Venezuela may follow a technocratic policy and China may adopt an ecodevelopmental policy in both the management of its environment and the transfer of environmental information. Taking Brazil, Mexico, Egypt and India, countries analyzed more deeply in the present work, it is possible to draw a comparison of some points concerning the policy followed by those nations.

In their comprehensive national policies, all these countries include development as their main goal. Strategies to achieve this goal vary somehow from one country to another: from a humane pattern to an illiberal style, from a social to an entirely economic model. Overall, while in Brazil, Egypt and Mexico development means economic growth demonstrated through higher industrial production, in India, development aims at a better standard of living, manifested by the eradication of poverty and the guarantee of employment for all. As mentioned before, this diversity may be explained by the alignment characteristic of the political economy adopted by governments of these countries: Brazil's right-wing government has embraced a capitalist pattern of industrialization and consumption; Mexico and Egypt have experienced the conflict between a socialist option and an actually adopted capitalist style of development, just a little better than Brazil; the socialist government of India, on the other hand, in spite of social inequalities associated with the system of castes and tribes, has opted to be faithful to the oriental tradition of detachment and follow the country's own cultural orientation concerning the selection of technology and the pattern of production and consumption.

The results of the actual alignment of these countries suggest that the assumption raised in chapter 1 on the political factors contributing to environmental degradation has a basis of truth, that is, a true socialist style is more likely to have concern for the environment than a capitalist one, and under an authoritarian government—regardless of its political alignment—the environment is not a priority at all.

Another consequence of the different styles of development adopted by those countries constitutes the level of autonomy achieved by them. India is becoming technologically viable (or self-reliant) through ecodevelopment, while Brazil, Mexico and Egypt are increasingly dependent, entangled by inappropriate technology. Out of the latter group of countries, the Brazilian and Egyptian governments are guided by a typical technocratic ideology. Concerning Mexico, research work, public education programs and various activities of some governmental bodies in the field of AT indicate that this country is searching for its technological viability and independence. For the time being, Mexico seems to be in the transition from a technocratic to an ecodevelopmental ideology.

Although Brazil, Mexico, Egypt and India have been considered up to now in this analysis from the viewpoint of their differences concerning their patterns of development and environmental management, they seem to present quite the same style of interrelation between the government and the population. With slight variations, in all those countries people are not given much chance to participate in governmental decisions, to have the first word when their own social destiny is under consideration or to be listened to about the use of natural resources. This situation pertains to underdevelopment, either because the adopted form of government is a dictatorship or because the paternalist style adopted does not admit that the masses can be competent. Concerning the environment, the population in these countries has successfully been forcing their participation in governmental decisions and it is hoped that a greater respect for the peoples' voices will become commonplace among the Third World governmental authorities in near future.

Another common facet of Brazil, Mexico, Egypt and India is the voluminous legislation concerning protection of the environment. Also common is the lack of condition to apply law as a consequence of poor technology or insufficient number of qualified personnel, mainly in relation to the large territorial area to be controlled.

As pointed out in chapter 4, an excessive secrecy is often associated with environmental information, mainly in developing countries. Allegedly for reasons of national security, manifestations of this syndrome have been met during the present research in various countries. In Brazil, secrecy was detected at INPE and at the Ministry of Navy. At INPE, it came up regarding satellite images of specific areas. At the Ministry of Navy, authorities considered it an insult to be questioned about their actions concerning marine pollution control; trying to dissuade this researcher, they mentioned "military apparatus" as their main instrument of control. In India, information has been denied at both the Ministry of Health and at the National Informatics Center of the Electronics Commission. Foreign researchers could have access neither to their premises nor to the information concerning their areas of action. In Egypt, the same paranoia has been demonstrated by the Central Agency for Public Mobilization and Statistics (CAPMS), where "they do not have time to waste with foreign researchers" and "everything that is not published is secret."

Control of information is certainly a challenging political problem to be considered by INFOTERRA, if its mission of facilitating the access of users to environmental information is to be taken seriously. In developed countries, where the organization of information into documents or data bases is satisfactory, access to sources can be easy and impersonal. However, in developing countries, people and organizations are still the main sources of environmental information; then, if secrecy barriers are interposed or if the "right person" is not met, getting knowledge about a given situation becomes a real puzzle, which sometimes invalidates an entire research study.

As far as the organization of environmental information is concerned, there are many overlapping areas of action among governmental organizations in all countries observed, mainly regarding water quantity and quality and sanitation. This waste of resources is inconsistent in countries where an urgent need exists to develop systems or to acquire monitoring equipment or laboratories likely to collect and process reliable information. A partial cause for the insufficiency of funds for the organization of environmental information is that this area is considered solely as a part of scientific and technological information in many developing nations. In the countries here analyzed, only the Indian government grants environmental information an independent status, indicated by the existence of a separate policy and budget and by the transfer of responsibility for strategies formulation from the Department of Science and Technology to the Department of Environment. In Brazil, efforts have been made by SEMA to work together with the Instituto Brasileiro de Informação em Ciência e Tecnologia (IBICT) and then participate in the decisions concerning environmental information. Results are so far very pale.

One last point to be considered about information on the local environment of Brazil, Mexico and Egypt is that, besides being scarce, literature is available only in the mother tongue of the respective country. India, where scientific papers, legislation and other governmental documents are often available in both Hindi and English, is an exception.

The points here considered are evidence of the fact that a free flow of information is still a fallacy at both national and international planes, and that behind this information plot lurks an unfair socioeconomic order.

NOTES

1. BRAZIL. Presidência da República. *Metas e bases para a ação do governo*. Brasília, 1970.

2. BRAZIL. Presidência da República. *I plano nacional de desenvolvimento (PND): 1972–1974*. Brasília, 1971.

3. BRAZIL. Presidência da República. *PBDCT, basic plan for scientific and technological development: 1973–74*. Brasília, 1973.

4. BRAZIL. Ministério das Relações Exteriores. "Telegrama no. 289, de 24.02.72, da DNU para a DELBRASONU." Brasília, MRE, 1972. (Unpublished.)

5. BRAZIL. Presidência da República. *II PND, II national development plan (1975–1979)*. Brasília, 1974.

6. BRAZIL. Presidência da República. *II PBDCT, II basic plan for scientific and technological development*. Brasília, 1976.

7. BRAZIL. Presidência da República. *III plano nacional de desenvolvimento: 1980–85*. Brasília, 1979.

8. BRAZIL. Presidência da República. *III PBDCT, III plano básico de desenvolvimento científico e tecnológico*. Brasília, 1979.

9. DICKSON, D. Brazil learns its ecological lessons—The hard way. *Nature*, 275: 684, Oct. 26, 1978.

10. BAIARDI, A. Desmatamamento: O caso da Amazônia brasileira. *Revista Brasileira de Tecnologia*, 14(2): 5–19, Mar./Apr. 1983.

11. GRAINGER, A. The state of the world's tropical forests. *The Ecologist*, 10(1–2): 6–54, Jan. 1980.

12. BRAZIL. Secretaria Especial do Meio Ambiente. *A atuação da Secretaria Especial do Meio Ambiente*. Brasília, 1980, pp. 35, 45.

13. DIÁRIO OFICIAL DA UNIÃO, Sept. 2, 1981.

14. MEXICO. Subsecretaria de Mejoramiento del Ambiente. *Informe nacional gobierno de México*. Nairobi, PNUMA, 1982, pp. 3–4.

15. MEXICO. Comisión Intersecretarial de Saneamiento Ambiental. *México: Diez años después de Estocolmo*. Nairobi, 1982, pp. 3–4, 68–69.

16. EGYPT. Ministry of Planning. *The five-year plan: 1978–1982*. Vol. 1: *The general strategy for economic and social development*. Cairo, 1977.

17. UNIDO. *Long-term prospects of industrial development in the Arab Republic of Egypt*. Algiers, Fifth Industrial Development Conference for Arab States, 1980, p. 48.

18. Ibid., p. 22.

19. INDIA. Planning Commission. *Sixth five-year plan: 1980–85*. New Delhi, 1981, pp. 17, 21, 32, 34, 318–20.

20. INDIA. National Committee on Science and Technology. *Draft science and technology plan*. New Delhi, 1973, vol. 2, p. 455.

21. Ibid., p. 435.

22. INDIA. *Sixth five year plan*, pp. 34, 76, 343.

23. GANDHI, I. *Mrs. Gandhi's address at plenary session of the International Conference on Human Environment*. New Delhi, Department of Environment, 1981, p. 7.

24. APPUKUTTAN, N. *National information system for science and technology (NISSAT)*. New Delhi, Department of Science and Technology, 1977, p. 2; and INDIA. *Draft science and technology plan*, p. 436.

25. Most of the data have been collected through personal contact or at least checked with a regional expert, when it was outside this researcher's own knowledge. Besides INFOTERRA's *International Directory of Sources*, the following sources have also been used:

· BRAZIL. SEMA. *Catálogo nacional das instituições que atuam na área do meio ambiente: 1981–82*. Brasília, 1982.
· *Directorio del medio ambiente en America Latina y el Caribe*. Santiago, CEPAL/CLADES, 1977.
· IBICT. *Fontes de Informação em meio ambiente no Brasil*. Brasília, 1983.

- TRZYNA, T. C., and COAN, E. V. *World directory of environmental organizations*. 2d ed. Claremont, Calif., Public Affairs Clearinghouse, 1976.
- UNEP. *Directory of institutions and individuals active in environmentally-sound and appropriate technologies*. Oxford, Pergamon Press, 1979.

6.

Does Any Light Gleam at the End of the Tunnel?

The analysis of the environmental discussion, at both national and international levels, shows that there is not common agreement about the solutions for the problems and not even about the priority actions to be taken. The arguments usually characterize individuals by their social class and nations according to their economic group. As far as individuals are concerned, it has been observed that, as their basic needs are satisfied, the quality of the environment is also included among their concerns. Going further, the environmental movement has been associated with the middle class, which gets the most from governmental measures in favor of cleaning and preserving the environment. Thus, W. Beckerman suggests this privilege should stop in order to allow the whole society to intervene in environmental choice.[1]

In spite of the lack of agreement, some possible ways for the solution of Third World environmental problems have been assumed to lie partly in new developmental patterns and life styles. On the other hand, environmental information has also been considered as a facilitator, since it is believed that a better informed society can also be more concerned for its environment.

NEW DEVELOPMENT ORIENTATION

World advancement is still underlined by the polarity of developed and backward nations, and the remaining international disagreements are caused basically by the interests and priorities of wealthy advanced countries colliding with those of the poor underdeveloped nations.

However, since the Stockholm Conference, some understanding between the two blocs have started, and the concept of environmental
deterioration was broadened to include, side by side, the pollution caused
by chemical agents and the pollution derived from poverty and related
socioeconomic factors. Both poor and developed nations have since
responded to UNEP's call, and national programs in defense of the environment are innumerable. However, the concept of a transnational
organization being able to manage the global environment is still uto-
pian. N. H. Jacoby, a defender of this idea, sees two dominant drawbacks—nationalism and each country's difference in priorities and
needs—and hopes for a future solution of global dimension.[2]

The observation of the present trends concerning developmental patterns and living styles leads to the expectation of a decrease in the existing burden on the global environment. Some authors also foresee
the enhancement of life quality in backward countries as either cause
or consequence of environmental improvement. C. Furtado[3],[4] bitterly
anticipates a future balance as to the physical environment resulting
from a strategy of the capitalist system in which privileges are granted
to a minority in the industrialized countries and sharing of world wealth
by the Third World is prevented. Under his assumption, sociopolitical
dynamics of nations would control the access by new countries to the
life style created and maintained by industrial capitalism in order to
avoid the collapse of the system. However, within such an unfair and
authoritarian atmosphere, C. Furtado can still foresee the emergence
of a new international power structure, where the Third World would
have a voice. This would come as a result of industrialized nations
searching for cooperation in order to solve problems derived from
technological pressure over the ecological frontier and from the inter-
nationalization of the economy.[5]

Reshaping the North-South relationship into a new basis of social
justice and peace is also the proposal of the already mentioned Brandt
Report, *North-South: A Programme for Survival*. It proposes interdependence and development as preconditions of human survival; for the
attainment of this goal, informing and educating the population about
the importance of international cooperation is of great importance.[6]

On the national scene, C. Furtado foretells of developing countries
being forced to entirely reshape their social structure into a more egalitarian society than the present capitalist model.[7] Defending a style of
development centered on human beings, J. Galtung *et al.* show high
income countries adopting alternative ways of life, such as those of

intermediate technology, health food, simple living groups, self-sufficiency movements, personal growth movements and experiments with self-management.[8] One could add to this observation that the same tendency exists in the Third World with regard to the higher classes, mainly in relation to the preference for frugal and natural eating. Both observations augur better access to natural resources (especially food) by the Third World poor and a better protected environment, as a consequence of more rational consumption by wealthy groups.

O. Sunkel[9] speculates on Latin America and an alternative pattern to replace the present style of development aimed solely at economic growth. Such an alternative should have the quality of life as the main goal of development and, then, emphasize the following elements: the environmental dimension, a fair income distribution among the population, the access to natural resources by all and the basis for the stability of the development process itself. Some of the strategies for change suggested by O. Sunkel are the adoption of a new consumption pattern favorable to the lower classes, the creation of legal and institutional structures consistent with the new development style and intense public participation in governmental planning. In order to make participative planning really effective, the author recommends mobilization of the population through improved systems of information, mass communication and education, which should work for the heightening of social consciousness with regard to the environment and development. Although discussed within a Latin American context, those suggestions can actually be extrapolated to the Third World as a whole.

ENVIRONMENTAL INFORMATION: PRESENT STATUS AND TRENDS

From both a user and a librarian standpoint, regardless of whether they work in a developed country or in the Third World, the transfer of environmental information is a complex job due to some special characteristics of the subject. These have been discussed by authors like M. L. Dosa,[10] A. N. Somerville,[11] J. S. Gardner[12] and J. Friedlander.[13] Among the features pointed out by those authors, the following are fundamental to illustrate the field:

• The cross disciplinarity of the field, which draws upon so many different segments that most people cannot even agree on the boundaries of the environmental sciences.

- Published literature about various aspects of the environment are dispersed among a wide variety of sources.
- Environmental data are often unpublished, and knowledge of their existence can only be grasped after a long or intense experience in the field.
- A large amount of information from both governmental and private organization sources is prevented from being available to the public since it is classified.
- Validity of the available data is quite dubious, since the methods of collection are either unknown or not standardized.
- Environmental data are vulnerable to manipulation by political pressures from the government or emotional appeal on the part of conservationists or pressure from economic groups.
- The volume of data and literature is huge.
- Indexing of documents included in either data bases or printed indexes is inefficient in providing users with ready access from an environmental point of view.
- Data and literature are prone to rapid obsolescence.
- There is a scarcity of socially based guidelines on what sources and categories of data should be available to whom and through which channels.
- Government information services are not standardized.
- Informal channels of communication have major importance in the field, but librarians are still not prepared to use them to their advantage.

Solutions to cope with these problems have been tried within each country and on the international plane, where UNEP plays the difficult role of coordinator, trying to establish a global network of environmental information. Notwithstanding all these efforts, much action is still needed, and information professionals should be aware of environmental management trends so that information services may actually support managers' actions and scientific research. In this concern, special attention should be paid to observations, such as the ones made by N. Lee, who puts forward six areas where more research is demanded from environmental scientists for them to achieve a better performance in the control of pollution: (a) the pattern of waste creation by society; (b) the technical efficiency and cost effectiveness of the available methods of waste control; (c) the pattern of waste diffusion through air, water or on land and how it is diluted or accumulated; (d) the levels of pollution concentration and its effect on humans and their environment; (e) the social costs of pollution; (f) the most appropriate criteria and organizational system to control pollution.[14]

From the perspective of information transfer, one could add that more research is also required about users' preferences and needs of environmental information. Of the few published studies on users of environmental information, the one performed by N. Lee,[14] and another by H. M. Ingram[15] are very useful to characterize the area. N. Lee specifically uncovers the information needs of personnel in the area of pollution control:

- Reliable data on wastes and pollution concentration levels, uniformly collected, synthesized and supplied on a time-series basis.
- A recently written authoritative survey or small collection of key surveys, likely to identify the main problem areas relevant to the discipline, to establish the technology appropriate for tackling those problems and to draw the state of the art in the subject area.
- A list of references appropriate to the problem area under consideration.
- A cheap current awareness service to update these references.
- A list of research organizations, research and workers in the specific environmental area of interest.[14]

H. M. Ingram, on the other hand, discusses the information channels used in making environmental decisions. In practice, decision makers focus upon a quite limited number of alternatives in making choices, and not all consequences of each alternative are, in fact, taken into account. This restricts their information needs to the main facts and data, relative to the direct and immediate effects of a decision. Information becomes relevant to decision makers when they are convinced that it will help them, or that they cannot afford to ignore it. As a result of this predisposition of decision makers, information overload is a likely problem. Environmental decision makers are presented by H. M. Ingram as being resistant to change, closely related to their past experience and reluctant to consider conflict-generating information which may complicate their decision. The author suggests that the factors affecting decision makers' choice of information are:

- The issue context, that is, information is only considered on the basis of a familiar (or fashionable) issue.
- The source of information supplied: the decision maker is more willing to use sources that are not controversial and are likely to lead him to a successful decision in political terms.
- Content of information: decision makers are particularly receptive to categories of information that justify and legitimize their decision-making process.

- Characteristics of decision makers' background and experience influence their ability to collect and assimilate data.
- Rules and regulations give legitimacy to information used by decision makers.
- Learning capacity: the intellectual and emotional capacity of decision makers to adjust to new sources of information and to feedback from decisions made.
- Timing: decision makers will be most receptive to new information during the "sorting out" phase of an emerging issue.

The author mentions defective communication between national and local levels as one of the main obstacles to a timely information flow for decision making.[15]

Even though developed countries have already achieved great advances in meeting users' needs, in the Third World most of the problems are still to be resolved by information professionals in association with environmentalists, and their solution should be a priority in the transfer of environmental information.

The Future of Environmental Information in the Third World

Two main features are called to the attention of an analyst who focuses on what lies ahead in the field of environmental information at the international level: the increasing volume of literature and data and how the advancement of technology may affect the area. Among all the above mentioned features, these are crucial if the overall situation of the Third World is taken into consideration.

As far as information volume is concerned, J. S. Gardner understands that it will become overwhelming very soon as a result of technological development in the area of data collection (for example, remote sensing and monitoring techniques and equipment) and because of the "environmental crisis." Then, he foresees the chaos of environmental information services under a gigantic volume of data and, consequently, the doomsday of environmental management as well.[16] To avoid this catastrophe, librarians dealing with environmental information should be very critical in assessing publications to be acquired. Identical measures should be taken when collecting and processing environmental quality data. Finally, regarding both publications and data, the dissemination of information should emphasize precision and objectivity in order to offer the right amount of information to the user.

As to the coverage, A. Lahiri suggests that a national system of environmental information should be able to answer questions pertaining to environmental status and trends, environmental technology, research and education and socioeconomic indicators, policy planning and legislative measures and judicial decisions.[17]

Information and communication technologies may help to ease the above mentioned difficulties, but there is also a suspicion that they could render the situation even more complex. In this concern, S. A. Wolpert suggests that, within the next generation, the pattern of information usage by managers and researchers may be dramatically affected by technological changes in the area of information systems, and services demanded by users will certainly be very sophisticated.[18] To add to the complexity of the scene, F. Lancaster warns in his futuristic report *Toward Paperless Information Systems* that a paperless society is rapidly coming, and that current information facilities should be prepared to embark upon this technological era. Otherwise, he foresees an apocalyptic information scenery for future societies.[19]

Speaking about environmental information specifically, D. Rubin *et al.* foresee optimistically, for the end of the eighties, that information technology will allow for better public knowledge on environmental problems. The subscriber of the ''Home Information Utility''—as he calls the public information system—will, from his office or residence, select directly and receive the information by terminal. Examples of resources which will be stored in the ''Home Information Utility'' are:

- A complete index of library information on any given environmental subject and the complete texts of books and articles.
- A list of pending government decisions, bills and meetings affecting the environment, with an indication of their status.
- Background information on any important event.
- The ability to monitor any important public meeting on a home television screen.
- Consumer information to permit environmentally informed purchasing decisions.
- The ability to register an opinion instantaneously in a local or national referendum about an environmental subject.[20]

M. Davis, however, is not so optimistic about the future of environmental information when faced with a new technological era. She anticipates a great variety of technological and information resources, formal and informal, which will not be equally shared, since "the information-rich will become richer, and the information-poor will not even suspect what they have missed."[21] This prospect of information starvation foreseen by M. Davis represents a major fear for developing countries, which causes a resentment about the scarcity of equipment, qualified personnel and funds to acquire both technology and ready-made information.

H. East and N. Belkin have been sensitive to the problem faced by developing countries regarding the paperless society. They concluded that it is unlikely for the poor nations to have an equal role in this new international electronic information network without undergoing internal social conflicts and substantial economic hardship or political compromise in their international relationships.[22]

To minimize such a critical situation, developing countries should congregate in regional groups, with help from international organizations (mainly UN bodies), and invest massively in the establishment in the Third World of centers dedicated to information transfer of environment-related subjects, especially appropriate technologies. Development starts from freedom, and, because of the lack of knowledge, resources and autonomy, backward countries have not been free to select in the international market what suits them best. As a result they are imposed on with expensive old-fashioned technology that degrades their environment and overburdens their economies. The proposed centers should be regional in character, be administered by interested countries and act as a "technology clearinghouse" in consonance with the science and technology policy of countries of the specific region.

The role to be played by librarians in this new era has been anticipated by some authors. A. Neelameghan and S. Seetharama mention two human capabilities which will be especially demanded in tomorrow's electronic system: conceptualization and imagination. The authors see the information worker as an important element to prevent further alienation of modern man caused by the deepening rupture between the individual and his environment.[23] Concerned with both environmental and technological development, R. Munn suggests that librarians of both developed and developing countries have many challenges and opportunities ahead, represented by the search for new solutions regarding more appropriate technologies.[24] This suggestion should be taken as an incentive for librarians to plunge themselves into the content of information (as researchers and subject specialists), instead of dealing only with the externalities of information (that is, forms of presentation, means of organization and so on), as has been the essence of traditional library work.

From the discussion of the overall questions of development, cooperation among developing countries themselves acquires a major relevance, since they have similar economies, close technological levels, common problems and research needs and a vast potential to be shared mainly at the regional level. Since the union of developing countries in blocs would improve their internal conditions and grant them power in the global relationship, it seems that the first step in this direction should be, at the present transitional phase, moving the main center of interest of their own information systems from the achievements of postindustrial societies to the knowledge of Third World needs, actual resources and foreseen potentialities. Surely, information on the advances of science and technology in developed countries, indispensable because it is a part of a single universal context, would always be needed whenever foreign knowledge comes to be more appropriate than indigenous know-how in satisfying specific local needs. Besides representing an important move toward self-reliance, this changing of focus could promote the recovery of the regional identity of developing countries as members of the Third World, now bewitched by the living style of industrial capitalist countries.

To make the North-South dialogue really effective, information systems should be associated with less conventional means of information transfer as an attempt to find channels of a broader and freer reach, which could cross the barriers imposed by political and economic groups represented by local government enactments, by local and foreign government institutions, by transnational organizations and by international news agencies.

Keeping in mind the present actors of the ecological movement in the Third World, it may be suggested that information professionals in this area should heavily use NGOs as their partners in collecting data, presenting the information to various audiences and disseminating information to the broadest group. Examples of active NGOs have been the Friends of the Earth (in its Third World branches), Environmental Liaison Center (Kenya), Kerala Sastra Sahithya Parishad (India), Center for Science and Environment (India), Environmental Studies Group (India), Fundação Brasileira para a Conservação da Natureza (Brazil), Sociedade Brasileira de Ecologia (Brazil), Associação Gaúcha de Proteção ao Ambiente Natural (Brazil) and Fundación Bariloche (Argentina). These and many others in every Third World nation should be considered by librarians as sources and channels in the environmental information transfer process.

Among NGOs, the church has played an important part in public awareness enhancement and in environmental education, as one can see from the activities of organizations such as the World Council of Churches (Switzerland); Conferência Nacional dos Bispos do Brasil (CNBB) in Brazil, mainly through its specific Pastorals and through the Comunidades Eclesiais de Base (CEB); Christian Action for Development in the Caribbean (Barbados); National Christian Council of Kenya; Christian Development Agency (Uganda) and Melanesian Council of Churches (Papua, New Guinea). The international network constituted by lay and religious NGOs seems to be the most dynamic and independent channel for the transfer of environmental information to the masses. They should be supported by international and national organizations and by all members of society who value information and freedom so that they can set up strong environmental information units for public use.

A second group to be looked for by information professionals in the transfer of environmental information to the population are those from the area of mass communication, since they have been very dynamic in the Third World with public mobilization and indirect control of government actions in relation to environment.

As a technological alternative for dissemination of environmental information, the media should be heavily used, together with the more conventional means of information transfer, since they are more likely to reach a larger audience than information systems alone. Newspapers and magazines, which are conventional forms of mass communication, should be used in the revolutionary way of an ''integral journalism,'' as recommended by A. Gramsci: not just meeting the needs of their specific public, but also creating new audiences and developing new demands among the public, within the frame of a progressive ideology devoted to changing social structure.[25] The alternative press has long been trying this line of action, and its leanings toward social nonconformance make it a natural channel for environmental news and campaigns. But, within the Third World's social reality, printed material discriminates against the majority of the population, since illiteracy prevails. Television and radio are then the most obvious communication means to channel the ecological message to the masses, and indeed they have an enormous audience.

Radio's power of penetration in the Third World has already been proved in the international political context of a specific region and as a national force to mobilize working classes in both urban and rural areas. During its short life of twenty months, Radio Notícias del Continente, created by the Federación Latinoamericana de Periodistas (FELAP) in Costa Rica in 1979, was an example that an independent radio can voice the reality of oppressed people to the world, in spite of local and foreign opposing forces. Another example of the use of radio for political purposes is the Egyptian Broadcasting Corporation, required by the government to strengthen national consciousness and to project the Arab world internationally.

In different countries, radio has also been largely used for rural communication. In Brazil, for instance, there is a vast network of radio stations, maintained by farmers' cooperatives or other organizations. However, telecommunication rights in Brazil are a property of the central government, which grants license of exploitation to private enterprises but still controls its use. The direct presence of government is probably an inhibitory factor for the social action that those means of rural communication could have. There are many radio stations operating in this context in Brazil, but the content of programs refer only to the agricultural policy of government, prices of agricultural products, meteorological information and countryside music. Ecologically sound technologies for agriculture, news about the real political and economic situation of the country and the rights of workers are not themes included in the programs. A space is open, however, for more aggressive professionals to seed environmental information throughout the country by using the already existent channels.

The prospects of hard work and intensive struggle in defense of the environment and in favor of a free flow of information among nations are such that information professionals should reevaluate their beliefs and strategies. Librarians have been blamed for being mentally rigid and socially apathetic. This static attitude librarians sometimes present to the public may be associated with the praxis of a traditional librarianship, characterized by conventionalism and rigid structures (the "mystery and mystique," criticized by M. Line[26]) and guided by introversion. Thankfully, this stereotype is giving way to a really social profession, more open to communication with the public and more concerned with its own space. The demands of modern society cannot be met by that aloof professional of old times anymore.

Summarizing the papers of the eighteenth Allerton Park Institute, H. Goldhor calls on librarians for deeper and more political participation in environmental issues. Such an attitude would imply a struggle for the free circulation of all existing information, against monopoly and censorship,[27] quite real ghosts of the environmental information panorama in the Third World.

Part of the librarians' political role should be their work for the emergence of a popular consciousness on the part of individuals about their rights as human beings and citizens in order to stop inequality and cultural disaggregation within society. P. Freire believes that the rise of public consciousness among the masses of a given dependent society prepares it to break the culture of silence imposed on the dominated by the dominant society.[28] Information and education are two possible roads leading to the rescue of dominated groups and then to the strengthening of civil society as a whole. The end result of such a process would be an environment dignified by the implementation of real national democracies and of fair world relationships.

Since the environmental crisis has its roots in socioeconomic problems, its solutions must be looked for beyond technological remedies for pollution of natural elements. From the standpoint of developing countries this solution should thus be sought in four directions:

- Through the establishment of a new international economic order, where every nation could have equal opportunities.
- In the minimization of poverty, inequalities, oppression and ignorance within developing societies, since it is expected that, when every individual in the Third World has at least the minimum condition to grow and develop his entire potentiality as a human being, the whole population will respond to the environmental appeal.
- In the increasing participation of civil society in governmental decisions through democratic institutions.
- In the adoption of nonpolluting technology, appropriate to the social and natural environment of each Third World nation and representative of alternatives freely chosen by society.

In spite of the large gaps between the present status and those envisaged goals, a profound hope in the future entitles the Third World to keep searching for environmental solutions through development.

However, for this move to be effective, the relationship among individuals and nations should be based on justice, and nations should work in total communion, since Earth is one, and individuals are a part of the same humankind. Either all will survive or all will together wreck under the negative forces built up by people's own selfishness. Environmental information can then play an important political role in guiding humans' reflection about their common habitat toward world solidarity of nations.

Utopia or a possible dream? It will depend on the attitude of every individual in the world. Vesting the spirit of Cervantes's Don Quixote de la Mancha, one could proclaim the belief that, while an individual's dream results in a chimera, a collective dream represents the dawn of reality.

P.S. to my colleague librarians: If we accept environmental information to be a possible way to a global solution, are we, information professionals, conscious that the light seemingly at the end of the tunnel may be in our own hands?

NOTES

1. BECKERMAN, W. *In defence of economic growth*. London, J. Cape, 1974, p. 248.

2. JACOBY, N. H. Organization for environmental management: National and transnational. *Management Science*, 19(10): 1138–50, June 1973.

3. FURTADO, C. Meio-ambiente, desenvolvimento e subdesenvolvimento na teoria econômica e no planejamento. In: ANDRADE, M. C. *et al. Meio-ambiente, desenvolvimento e subdesenvolvimento*. São Paulo, Hucitec, 1975, pp. 83–90.

4. FURTADO, C. *O mito do desenvolvimento econômico*. 5th ed. Rio de Janeiro, Paz e Terra, 1981, pp. 68–76.

5. FURTADO, C. *O Brasil pós-"milagre."* 3rd ed. São Paulo, Paz e Terra, 1981, p. 114.

6. BRANDT, W. From the Brandt Report. *IDR*, 4: 88–103, 1980.

7. FURTADO, C. Meio-ambiente, pp. 89–90.

8. GALTUNG, J.; PREISWERK, R.; and WEMEGAH, M. A concept of development centered on the human being: Some Western European perspectives. *Canadian Journal of Development Studies*, 2(1): 145–46, 1981.

9. SUNKEL, O. *La dimensión ambiental en os estilos de desarrollo de América Latina*. Santiago de Chile, CEPAL, 1981, pp. 132–35.

10. DOSA, M. L. Reflections on the workshop. In: DAVIS, M. (ed.). *En-

vironmental information: Some problems and solutions. Melbourne, Center for Environmental Studies, Melbourne University, 1974, p. 23.

11. SOMERVILLE, A. N. Academia and the environment: Environmental information. *Journal of Chemical Information and Computer Sciences*, 16(1): 1–4, 1976.

12. GARDNER, J. S. *A study of environmental monitoring and information systems*. Iowa City, Iowa University, 1972, p. 11.

13. FRIEDLANDER, J. Coping with environmental information resources. In: BONN, G. S. (ed.). *Information resources in the environmental sciences*. Papers presented at the eighteenth Allerton Park Institute, November 12–15, 1972. Urbana-Champaign, University of Illinois, Graduate School of Library Science, 1973, pp. 2003–4.

14. LEE, N. The information requirements of the research worker. In: ASLIB and LIBRARY ASSOCIATION. *Pollution: Sources of information*. London, 1972, pp. 12–19.

15. INGRAM, H. M. Information channels and environmental decision making. *Natural Resources Journal*, 13(1): 150–69, Jan. 1973.

16. GARDNER, J. S. *A study of environmental monitoring and information systems*, pp. 11, 259.

17. LAHIRI, A. Towards an information system on environment. *Library Science*, 15(1): 10, Mar. 1978.

18. WOLPERT, S. A. The information manager of the future. In: AMERICAN SOCIETY FOR INFORMATION SCIENCE. *Proceedings of the ASIS annual meeting*. Sept. 26–Oct. 1, 1977. Chicago, 1977, vol. 14, pt. 2, pp. 1–7.

19. LANCASTER, F. W. *Toward paperless information systems*. New York, Academic Press, 1978, p. 166.

20. RUBIN, D. M. *et al*. Environmental information. *Annals of New York Academy of Sciences*, 216(7): 175–76, 1973.

21. DAVIS, M. (ed.). *Environmental information: Some problems and solutions*. Melbourne, Center for Environmental Studies, Melbourne University, 1974, p. 27.

22. EAST, H., and BELKIN, N. J. *Advanced technology and the developing countries: The growing gap*. Edinburgh, FID Conference, 1978.

23. NEELAMEGHAN, A., and SEETHARAMA, S. Information transfer: The next twenty-five years. *Library Science*, 13(1): 11–12, Mar. 1976.

24. MUNN, R. F. Appropriate technology and information services in developing countries. *International Librarianship Review*, 10: 23–27, 1978.

25. GRAMSCI, A. *Os intelectuais e a organização da cultura*. Rio de Janeiro, Civilização Brasileira, 1979, pp. 161–63.

26. LINE, M. B. Demystification in librarianship and information science. In: BARR, K. and LINE, M. *Essays on information and libraries*. London, C. Bingley, 1975, pp. 107–16.

27. GOLDHOR, H. A summary and overview of the conference. In: BONN, G. S. (ed.). *Information resources in the environmental sciences*. Papers presented at the eighteenth Allerton Park Institute, November 12–15, 1972. Urbana-Champaign, University of Illinois, Graduate School of Library Science, 1973.

28. FREIRE, P. *Conscientização: Teoria e prática da libertação—Uma introdução ao pensamento de Paulo Freire*. São Paulo, Cortez & Moraes, 1980, p. 65.

Epilogue

In travelling to the state capitals of Brazil and to many developing countries to gather data for the present research, I have acquired a new perception of environmental information as a human fraternity. Even though a few authorities opted for a noncommunication posture, many others in both the public and the private sectors have shared their knowledge and their warmth.

Through this experience I also came to understand that the individual himself, as a representative of a specific culture, is an essential dimension of environmental information. Thus, everyone in the street seemed to me to be a witness to the local context and a testimony to his time and environment.

This human perspective of environmental information is an open door to research and library work: the saga of the human environment is there to be written, and every individual has a piece of information to contribute to it. On the other hand, the rescue of millions of individuals from underdevelopment demands from librarians—as from all professionals—a deep involvement in society's goals and projects.

The hospitality of those men and women who have welcomed me, an unknown researcher, with the friendly attitude of serious professionals gave me the firm conviction that the cause of environmental information in developing countries has many allies, and that world solidarity is a very possible dream.

Main Institutional Sources on Environmental Information in the Third World

ORGANIZATIONS OF INTERNATIONAL SCOPE

International Development Research Center (IDRC)
P. O. Box 8500
Ottawa, Ontario KIG 3H9
Canada

International Reference Center for Community Water Supply
P. O. Box 140
2260 AC Leidschendam
Netherlands

World Council of Churches
150, Route de Ferney
1211 Geneva
Switzerland

Friends of the Earth
9 Poland St.
London WIV 3DC
England

Intermediate Technology Development Group Ltd. (ITDG)
9 King St.
Covent Garden
London WC2E SHN
England

Centro Internacional de Formación en Ciencias Ambientales (CIFCA)
Oficina de Documentación y Difusión
Serrano, 23 10 Piso
Madrid 1
Spain

Environmental Liaison Center
P. O. Box 72.461
Nairobi
Kenya

UN Food and Agriculture Organization (FAO)
Via delle Terme di Caracalla
00100 Rome
Italy

UN Center for Human Settlements
P. O. Box 30.030
Nairobi
Kenya

International Union for Conservation of Nature and Natural Resources
1196 Gland
Switzerland

UN Educational, Scientific and Cultural Organization (UNESCO)
Place de Fontenoy, 75
Paris 7
France

UN Environment Program (UNEP)
P. O. Box 30.552
Nairobi
Kenya
(Helpful contacts: Mr. T. Munetic and Mr. Kevin Grose)

UN Statistical Office
New York
United States
(Helpful contact: Mr. Peter Bartelmus)

USAID Environmental Affairs Coordination
U.S. International Development Cooperation Agency
21 St.
State Department Bldg.
Washington, D.C. 20523
United States
(Helpful contact: Mr. Albert C. Printz, Jr.)

World Bank
1818 H St., N.W.
Washington, D.C. 20433
United States
(Helpful contact: Dr. Robert Goodland)

World Health Organization (WHO)
Ave. Appia, 1211
Geneva 27
Switzerland

LATIN AMERICAN ORGANIZATIONS

Centro Panamericano de Ecología Humana y Salud (ECO)
Apartado Postal 249
Toluca
Edo. Mexico
Mexico

Centro de Información y Referencia en Ingeniería Sanitaria y Control del
    Ambiente (CIRISCA)
Blanco Encalada 2120, 4° Piso
Santiago
Chile

Centro Latinoamericano de Demografia (CELADE)
Casilla 91
Santiago
Chile

Centro Panamericano de Información Documentación en Ingeniería Sanitaria
    y Ciencias Ambientales (CEPIS)
Casilla 4337
Lima 100
Peru

Centro Latinoamericano de Documentación Económica y Social (CLADES)
Casilla 179–D
Santiago
Chile

CEPAL, Asentamientos Humanos
Av. Mazaryk 29, 40 Piso
Colonia Polanco
Mexico, 5, DF
Mexico
(Helpful contact: Mr. Eduardo Neira Alva)

Instituto de Desarrollo de los Recursos Naturales Renovables (INDERENA)
Carrera 14, no. 25, A, 66
Bogotá
Colombia

Pan American Health Organization
525, 23 Street, N.W.
Washington, D.C. 20037
United States
(Helpful contact: Dr. Guillermo Davilla)

Red Latinoamericana de Información sobre Asentamientos Humanos
  (LATINAH)
Apartado Aéreo 34.219
Bogotá
Colombia

Red Panamericana de Información y Documentación en Ingeniería Sanitaria y
  Ciencias del Ambiente (REPIDISCA)
Casilla 4337
Lima 100
Peru

Sistema de Documentación sobre Populación en America Latina (DOCPAL)
Casilla 91
Santiago
Chile

Sistema de Información para Planificación en America Latina y el Caribe
  (INFOPLAN)
Casilla 179–D
Santiago
Chile

UNEP Regional Office for Latin America
Av. Mazaryk 29
Colonia Polanco
Mexico, 5, DF
Mexico
(Helpful contact:Mr. Jaime Hurtubia)

BRAZILIAN ORGANIZATIONS

Associação Brasileira de Engenharia Sanitária e Ambiental (ABES)
C. P. 15.029
20.030 Rio de Janeiro, RJ
Brazil

Associação Gaúcha de Proteção ao Ambiente Natural (AGAPAN)
R. Jacinto Gomes, 39
90.000 Porto Alegre, RS
Brazil
(Possible contact: Mr. José A. Lutzenberger)

Centro de Pesquisas e Desenvolvimento (CEPED)
C. P. 09
48.240 Camaçari, BA
Brazil
(Helpful contact: Ms. Marilene Lobo Abreu Barbosa)

Companhia de Tecnologia de Saneamento Ambiental (CETESB)
Av. Prof. Frederico Herman, Jr. 345
05.459 São Paulo, SP
Brazil

Conselho Nacional do Desenvolvimento Cientifico e Tecnológico (CNPq)
Av. W3 N, Q. 507, Bl. B
70.740 Brasília, DF
Brazil
(Helpful contacts: Dr. Antonio Dantas Machado and Mr. Luiz Antonio
    Gonçalves)

Fundação Brasileira para a Conservação da Natureza (FBCN)
Praia do Botafogo, 210, S. 805/808
22.250 Rio de Janeiro, RJ
Brazil
(Helpful contact: Dr. José Cândido de Carvalho)

Fundação Centro Tecnológico de Minas Gerais (CETEC)
Av. José Cândido da Silveira, 2.000
Horto
30.000 Belo Horizonte, MG
Brazil
(Helpful contact: Ms. Maria Regina Gonçalves dos Santos)

Empresa Brasileira de Pesquisa Agropecuária (EMBRAPA)
Ed. Venâncio 2.000 - 60 e 90 andares
70.333 Brasília, DF
Brazil
(Helpful contact: Mr. Ubaldino Dantas Machado)

Fundação Estadual de Engenharia do Meio Ambiente (FEEMA)
Rua Fonseca Teles, 121, 150 andar, São Cristóvão
20.940 Rio de Janeiro, RJ
Brazil

Instituto Brasileiro de Desenvolvimento Florestal (IBDF)
SAIN, Av. L4
70.910 Brasília, DF
Brazil

Instituto Brasileiro de Informação em Ciência e Tecnologia (IBICT)
SAS.Q5, L.6, Bl. H
70.000 Brasília, DF
(Helpful contact: Ms. Yone S. Chastinet)

Instituto Nacional de Tecnologia (INT)
Av. Venezuela 82
20.081 Rio de Janeiro, RJ
Brazil

Secretaria de Tecnologia Industrial (STI)
Ministério da Indústria e Comércio
SAS Q. 2, L. 3, S. 1
Ed. INPI
70.053 Brasília, DF
Brazil
(Helpful contact: Mr. José Rincon Ferreira)

Secretaria Especial de Meio Ambiente (SEMA)
Av. W3N, Q.510
Esplanada dos Ministérios
Ed. Cidade de Cabo Frio
Brazil
(Helpful contact: Mr. Estanislau Monteiro de Oliveira)

Sistema de Documentação sobre População no Brasil. Fundação Sistema
    Estadual de Análise de Dados (SEADE)
C. P. 8.223
01033 São Paulo, SP
Brazil

Superintendência de Recursos Naturais e Meio Ambiente da FIBGE
R. Equador, 558, 2° andar, Santo Cristo
20.220 Rio de Janeiro, RJ
Brazil

MEXICAN ORGANIZATIONS

Centro de Ecodesarrollo
Chilaque 41

Colonia Churubusco
Mexico City, 21, DF
Mexico
(Helpful contact: Mr. Alejandro Toledo)

Centro de Estudios del Meio Ambiente
Universidad Autónoma Metropolitana
Azcopotzalco, A.P. 16–306
Mexico, DF
Mexico

Centro de Investigaciones Científicas y Tecnologicas
Instituto Nacional de Investigación sobre Recursos Bióticos (INEREB)
Pedregal de San Angel
Delegación Alvaro Obregon
01.900 Mexico, DF
Mexico

Consejo Nacional de Ciencia y Tecnología (CONACYT)
Insurgentes Sur 1677, P. B.
Mexico, 20, DF
Mexico

Instituto de Ciencias del Mar y Liminología
A. P. 811
Mazatlan, Sinaloa
Mexico

Instituto Mexicano del Petroleo
Av. de los Cién Metros, 152
Mexico, DF
Mexico

Secretaria de Agricultura y Recursos Hidraulicos (SARH)
Insurgentes Sur, 540, 70 Piso
Mexico, DF
Mexico
(Helpful contact: Eng. Francisco Bahamond Torres)

Secretaria de Asentamientos Humanos y Obras Publicas (SAHOP)
Av. Constituyentes 947
Ed. B, P. B.
Mexico, 18, DF
Mexico
(Helpful contacts: Eng. Luiz Sanchez de Carmona and Mr. Federico Lopez
    de Alba)

Secretaria de Programación y Presupuesto Coordenación General de los
    Servícios Nacionales de Estatística, Geografia e Informática
Balderas, 71
Mexico, DF
Mexico

Secretaria de Relaciones Exteriores
Departamiento de Medio Ambiente
Av. Ricardo Flores Magón, 1
Mexico City, 1, DF
Mexico
(Helpful contact: Ms. Adriana Aguilera)

Subsecretaria de Mejoramiento del Ambiente
Av. Chapultepec, 284
Mexico, 7, DF
Mexico
(Helpful contacts: Dr. Fidel Mascareño Sauceda and Dr. Guilhermo Diaz
    Mejía)

AFRICAN ORGANIZATIONS

Centre de Coordination des Recherches et de la Documentation en Sciences
    Sociales Desservant l'Afrique Subsaharienne (CERDAS)
Kinshasa
Zaire

Inter-African Bureau of Soils
Organization of African Unity (OAU/STRC)
B. P. 1352
Bangui
Central African Republic

Science and Technology Research Center
Organization of African Unity (OAU)
P. O. Box 2359
Lagos
Nigeria

Science and Technology Unit
UN Economic Commission for Africa (ECA)
P. O. Box 3001
Addis Ababa
Ethiopia

UN Information Center
Harambee Ave.
Box 30 218
Nairobi
Kenya

EGYPTIAN ORGANIZATIONS

Academy of Scientific Research and Technology
101 Kasr El-Aine St.
Cairo
Egypt
(Possible contacts: Dr. Mohamed Al-Kassás and Ms. Meguid)

Central Agency for Public Mobilization and Statistics (CAPMAS)
P. O. Box 2086
Nasr City
Cairo
Egypt
(Director at the time of this research: Dr. Haluda)

General Organization for Industrialization (GOFI)
6 Khalil Agha St.
Garden City
Cairo
Egypt
(Helpful contact: Eng. Kamel Maksoud)

High Institute of Public Health
Alexandria University
165 El-Horria Ave.
Alexandria
Egypt
(Helpful contacts: Dr. Ahmed Hamza and Dr. Samia Galal Saad)

Ministry of Health
Cairo
Egypt
(Helpful contact: Dr. Gama El-Samra)

National Information and Documentation Center (NIDOC)
El-Tahir St., Dokki
Giza

Egypt
(Helpful contact: Dr. Aadl Duweini)

Office of the Transfer of Technology
El-Tahir St., Dokki
Giza
Egypt
(Helpful contact: Prof. Mahmond Badel Halim Saleh)

Water Pollution Control Laboratory
National Research Center
El-Tahir St., Dokki
Giza
Egypt
(Helpful contact: Dr. Fatma Gohary)

ASIAN ORGANIZATIONS

Asian Institute of Technology
P. O. Box 2754
Bangkok
Thailand

Development Technology Center
Institute of Technology Bandung
Jalan Ganesha 10
P. O. Box 276
Bandung
Indonesia

Regional Center for Environment and Resources Study for Asia and the
    Pacific
Faculty of Environment and Resources Studies
Mahidol University
Dhonburi
Bangkok, 7
Thailand

UN Economic and Social Commission for Asia and the Pacific (ESCAP)
UN Building
Rajadamnern Ave.
Bangkok, 2
Thailand

INDIAN ORGANIZATIONS

Central Board for the Prevention and Control of Water Pollution
Skylam Building, 5th Floor
60 Nehru Place
New Delhi
India
(Helpful contact: Mr. H. S. Matharu)

Central Public Health and Environmental Engineering Organization
Ministry of Works and Housing
Nirman Bhawan
New Delhi, 110.011
India
(Helpful contact: Mr. M. K. Moitra)

Central Statistical Organization
Ministry of Planning
Sarder Patel Bhawan
Parliament Street
New Delhi, 110.001
India

Center for Science and Environment (CSE)
807 Vishal Bhawan
95 Nehru Place
New Delhi, 110.019
India
(Helpful contact: Mr. Anil Agarwal)

Center of Social Medicine and Community Health
Jawaharlal Nehru University (JNU)
Old Campus, Bl. 3
New Delhi, 110.057
India
(Helpful contact: Dr. D. Banerjee)

Council of Scientific and Industrial Research (CSIR)
Rafi Marg
New Delhi, 110.001
India

Department of Environment
Bikaner House
Shanjahan Road
New Delhi, 110.011

India
(Helpful contacts: Dr. A. Khosla and M.P.R. Banerjee)

Department of Science and Technology
Technology Bhawan
New Mehrauli Road
New Delhi, 110.029
India
(Helpful contacts: Dr. A. Lahiri and Mr. Harjit Singh)

Environment Service Group
WWF, India
39 Uday Park
New Delhi, 110.049
India
(Helpful contact: Dr. T. Mathew)

Indian Institute of Science
Bangalore, 560.012
India

Indian National Scientific Documentation Center (INSDOC)
Hillside Road
New Delhi, 12
India

Indian Statistics Institute
Bangalore, 560.003
India

Kerala Sastra Sahitya Parishad (KSSP)
Parishad Bhawan
Chirakulan Road
Trivandrum, 695.001
India

National Informatics Center
Electronics Commission
E Wing, Pushpa Bhawan
Madangir Road
New Delhi, 62
India
(Director at the time of this research: Dr. S. N. Seshagiri)

National Environmental Engineering Research Institute (NEERI)
Nehru Marg
Nagpur, 440.020
India
(Possible contact: Dr. B. B. Sundaresan)

National Research Development Corporation (NRDC)
61 Ring Road
New Delhi, 110.024
India

School of Environmental Sciences
Jawaharlal Nehru University
New Campus
New Delhi, 110.057
India
(Helpful contact: Professor J. Dave)

Urban Development
Ministry of Works and Housing
Nirman Bhawan
New Delhi, 110.011
India
(Helpful contact: Mr. P.S.A. Sundaran)

Bibliography

ALMEIDA, M. O. The confrontation between problems of development and environment. *International Conciliation*, Jan. 1972, special issue, environment and development.

ALTHUSSER, L. *Ideologia e aparelhos ideológicos de estado*. Lisboa, Presença, 1980.

AMERICAN SOCIETY FOR INFORMATION SCIENCE. *Proceedings of the ASIS annual meeting*. Sept. 26–Oct. 1, 1977. Chicago, 1977, vol. 14.

ANDRADE, M. C., *et al. Meio-ambiente, desenvolvimento e subdesenvolvimento*. São Paulo, Hucitec, 1975.

APPUKUTTAN, N. *National information system for science and technology (NISSAT)*. New Delhi, Department of Science and Technology, 1977.

ASLIB and LIBRARY ASSOCIATION. *Pollution: Sources of information*. London, 1972.

BAIARDI, A. Desmatamamento: O caso da Amazônia brasileira. *Revista brasileira de Tecnologia*, 14(2): 5–19, Mar./Apr. 1983.

BARAN, P. A. *The political economy of growth*. Middlesex, England, Penguin Books, 1973.

BARR, K., and LINE, M. *Essays on information and libraries*. London, C. Bingley, 1975.

BASSOW, W. Third World is "going environmental." *The Journal of Commerce and Commercial*, New York, Dec. 30, 1980.

BECKERMAN, W. *In defence of economic growth*. London, J. Cape, 1974.
———. *Two cheers for the affluent society*. London, St. Martin, 1976.

BELL, G., and TYRWHITT, J. *Human identity in the urban environment*. Middlesex, England, Penguin, 1972.

BERNSTEIN, H. (ed.). *Underdevelopment and development: The Third World today*. Middlesex, England, Penguin, 1973.

BISWAS, A. K. Estudios globales futuros: Una revision de la decada pasada. *Mazingira*, 6(1): 68–75, 1982.

BLANC, G. The world of appropriate technology. *Appropriate Technology*, 9(4): 14–15, Mar. 1983.

BOLAFFI, G. A questão urbana, produção de habitações, construção civil e mercado de trabalho. *Novos Estudos Cebrap*, 2(1): 65, Apr. 1983.

BONN, G. S. (ed.). *Information resources in the environmental sciences*. Papers presented at the eighteenth Allerton Park Institute, November 12–15, 1972. Urbana-Champaign, University of Illinois, Graduate School of Library Science, 1973.

BONO, E. *Ecologia e política á luz do Tao*. Porto Alegre, Brazil, Record, 1982.

BORGSTROM, G. *Too many*. New York, Macmillan, 1969.

BOWMAN, J. S. Environmental coverage in the mass media: A longitudinal study. *International Journal of Environmental Studies*, 18: 11–22, 1981.

BRANDT, W. From the Brandt Report. *IDR*, 4: 88–103, 1980.

BRAZIL. Ministério das Relações Exteriores. "Telegrama no. 289, de 24.02.72, da DNU para a DELBRASONU." Brasília, MRE, 1972. (Unpublished.)

————. Presidência da República. *Metas e bases para a ação do governo*. Brasília, 1970.

————. Presidência da República. *PBDCT, basic plan for scientific and technological development: 1973–74*. Brasília, 1973.

————. Presidência da República. *I plano nacional de desenvolvimento (PND): 1972–1974*. Brasília, 1971.

————. Presidência da República. *II PBDCT, II basic plan for scientific and technological development*. Brasília, 1976.

————. Presidência da República. *II PND, II national development plan (1975–1979)*. Brasília, 1974.

————. Presidência da República. *III PBDCT, III plano básico de desenvolvimento científico e tecnológico*. Brasília, 1979.

————. Presidência da República. *III plano nacional de desenvolvimento: 1980–85*. Brasília, 1979.

————. Secretaria Especial do Meio Ambiente. *A atuação da Secretaria Especial do Meio Ambiente*. Brasília, 1980.

————. Secretaria Especial do Meio Ambiente. *Catálogo nacional das instituições que atuam na área do meio ambiente: 1981–82*. Brasília, 1982.

————. Secretaria Especial do Meio Ambiente. *Meio ambiente no Brasil: Evolução e perspectiva histórica*. Brasília, 1982.

BRESSER PEREIRA, L. C. Semiverdades e falsas idéias sobre o Brasil. *Novos Estudos Cebrap*, 2(2): 24, July 1983.

BROWN, L. R. *Building a sustainable society*. New York, W. W. Norton, 1981.

BROWN, L. R., and SHAW, P. *Six steps to a sustainable society*. Washington, D.C., Worldwatch Institute, 1982.

CAMPBELL, A., *et al. The quality of American life*. New York, Russell Sage Foundation, 1976.

CARSON, R. *Silent spring*. London, H. Hamilton, 1962.

CASTRO, J. de. Subdesenvolvimento, causa primeira da poluição. *O Correio da UNESCO*, 1(3): 20, 1973.

CENTER FOR SCIENCE AND ENVIRONMENT. *The state of India's environment, 1982: A citizens' report*. New Delhi, 1982.

CHILCOTE, R. H. Teorias reformistas e revolucionárias de desenvolvimento e subdesenvolvimento. *Revista de Economia Política*, 3(3): 103–23, July/Sept. 1983.

CHOMSKY, N., and HERMAN, E. S. *The political economy of human rights*. London, Spokesman, 1979. Vol. 1: *The Washington connection and the Third World fascism*. Vol. 2: After the cataclysm.

CLAXTON, R. Ambientalismo latino-americano: Fraco e limitado às elites. *Raízes*, 2: Aug. 1977.

COMMISSION OF THE EUROPEAN COMMUNITIES. Continuation and implementation of a European Community policy and action programme on the environment. *Bulletin of the European Communities*, Supplement 6/76.

COMMONER, B. *Closing circle: The environmenal crisis and its cure*. London, J. Cape, 1972.

———. *Science and survival*. New York, Viking, 1967.

COULSTON, F., and KORTE, F. *Environmental quality and safety*. Stuttgart, G. Thieme, 1976.

COX, P. R., and PEEL, J. (eds.). *Population and pollution*. London, Academic Press, 1972.

DAGNINO, R. P. Indústria de armamentos: O estado e a tecnologia. *Revista Brasileira de Tecnologia*, 14(3): 6, May/June 1983.

DAVIS, M. (ed.). *Environmental information: Some problems and solutions*. Melbourne University, 1974.

DIÁRIO OFICIAL DA UNIÃO. Sept. 2, 1981.

DICKSON, D. Brazil learns its ecological lessons—The hard way. *Nature*, 275: 684, Oct. 26, 1978.

Directorio del medio ambiente en America Latina y el Caribe. Santiago, CEPAL/CLADES, 1977.

DORST, J. *Before nature dies*. London, Collins, 1971.

DUBOS, R. *Man adapting*. New Haven, Conn. Yale University Press, 1967.

———. *So human an animal*. New York, Scribner's, 1968.

EAST, H., and BELKIN, N. J. *Advanced technology and the developing countries: The growing gap*. Edinburgh, FID Conference, 1978.

EBERSTADT, N. Fertility declines in less-developed countries: Components and implications. *Environmental Conservation*, 8(3): 187–89, Autumn 1981.

EFFROTT, M. P. (ed.). *The community approaches and applications*. New York, Free Press, 1974.

EGYPT. Ministry of Planning. *The five-year plan: 1978–1982*. Vol. 1: *The general strategy for economic and social development*. Cairo, 1977.

EHRLICH, A. H., and EHRLICH, P. R. Dangers of uninformed optimism. *Environmental Conservation*, 8(3): 173–75, Autumn 1981.

EHRLICH, P. *Population bomb*. San Francisco, Ballantine, 1968.

EHRLICH P. and EHRLICH, A. H. *Population, resources, environment: Issues in human ecology*. San Francisco, W. H. Freeman, 1970.

EHRLICH, P., and HARRIMAN, R. L. *How to be a survivor*. London, Ballantine, 1971.

EHRLICH, P., and HOLDREN, J. P. *Global ecology: Readings toward a rational strategy for man*. New York, Harcourt, Brace, Jovanovich, 1971.

EL-HINNAWI, E. El medio ambiente mundial: ¿Ahora, hacia donde? *Mazingira*, 6(1): 62, 1982.

EL-SHAKHS, S. The population bomb and urbanization. *Ambio*, 12(2): 94–96, Apr. 1983.

FADIA, B. *Pressure groups in Indian politics*. New Delhi, Radiant, 1980.

FINSTERBUSCH, K. Consequences of increasing scarcity on affluent countries. *Technological Forecasting and Social Change*, 23: 59–73, 1983.

FRANK, A. G. *Capitalism and underdevelopment in Latin America: Historical studies of Chile and Brazil*. Middlesex, England, Penguin, 1969.

———. Desenvolvimento do subdesenvolvimento latino-americano. In: PEREIRA, L. *Urbanização e subdesenvolvimento*. 4th ed. Rio de Janeiro, Zahar, 1979.

FREIRE, P. *Conscientização: Teoria e prática da libertação—Uma introdução ao pensamento de Paulo Freire*. São Paulo, Cortez & Moraes, 1980.

———. *Educação e mudança*. São Paulo, Paz e Terra, 1981.

FROMM, E. *To have or to be*. London, Sphere Books, 1979.

FUNDACIÓN BARILOCHE. *Modelo mundial latinoamericano: Informe presentado en el International Institute for Applied Systems Analysis*, Vienna, Oct. 1974.

FURTADO, C. *O Brasil pós-"milagre."* 3d ed. São Paulo, Paz e Terra, 1981.

———. *O mito do desenvolvimento econômico*. 5th ed. Rio de Janeiro, Paz e Terra, 1981.

GALTUNG, J.; PREISWERK, R.; and WEMEGAH, M. A concept of development centered on the human being: Some Western European perspectives. *Canadian Journal of Development Studies*, 2(1): 145–46, 1981.

GANDHI, I. *Mrs. Gandhi's address at plenary session [of the International*

Conference on Human Environment]. New Delhi, Department of Environment, 1981.

GARDNER, J. S. *A study of environmental monitoring and information systems*. Iowa City, Iowa University, 1972.

GERELLI, E., and BARDE, J. P. *A qui profite l'environment, riches ou pauvres?* Paris, Presses Universitaires de France, 1977.

GERRY, C. Restructuring underdevelopment? *Centre for Development Studies Newsletter*, 5: 3–7, Dec. 1981.

GONZÁLEZ CASANOVA, P. and FLORESCANO, E. *México, hoy*. 6th ed. México City, Siglo Veintiuno, 1982.

GRAINGER, A. The state of the world's tropical forests. *The Ecologist*, 10(1–2): 6–54, Jan. 1980.

GRAMSCI, A. *Os intelectuais e a organização da cultura*. Rio de Janeiro, Civilização Brasileira, 1979.

GUERREIRO, R. S. A crise existe tanto no Sul como no Norte. *Jornal do Brasil*, Caderno Especial, p. 3, Nov. 1, 1981.

HARDOY, J., and SATTERTHWAITE, D. *Shelter: Need and response—Land and settlement policies in 17 Third World nations*. New York, J. Wiley, 1981.

HECOX, W. E. Limits to growth revisited: Has the world modelling debate made any progress? *Environmental Affairs*, 5(1): 65–96, Winter 1976.

HINES, W. W., and WILLEKE, G. E. Public perceptions of water quality in a metropolitan area. *Water Resources Bulletin*, 10(4): 745–55, Aug. 1974.

HOLY BIBLE. O. T. *Genesis*, chap. 1, ver. 28–29. Catholic edition. New York, Nelson, 1966.

IBICT. *Fontes de Informação em meio ambiente no Brasil*. Brasília, IBICT, 1983.

IKRAM, K. *Egypt; Economic management in a period of transition*. Baltimore, Md., Johns Hopkins University, 1980 (A World Bank Country Economic Report).

INDIA. National Committee on Science and Technology. *Draft science and technology plan*. New Delhi, 1973.

————. Planning Commission. *Sixth five year plan: 1980–85*. New Delhi, 1981.

INGRAM, H. M. Information channels and environmental decision making. *Natural Resources Journal*, 13(1): 150–69, January 1973.

IOHANNES PAULUS II, P. P. *Redemptor hominis*. Rome, 1979.

IUCN. *World conservation strategy: Living resource conservation for sustainable development*. Gland, Switzerland, IUCN, 1980.

JACOBY, N. H. Organization for environmental management: National and transnational. *Management Science*, 19(10): 1138–50, June 1973.

KAHN, H. Our global growing pains. *Nation's Business*, pp. 32–38, July 1973.

KOCHEN, M. (ed.). *Information for action: From knowledge to wisdom.* New York, Academic Press, 1975.

KRANZBERG, M. Technology and human values. *Dialogue*, 11(4): 21–29, 1978.

LAHIRI, A. Towards an information system on environment. *Library Science*, 15(1): 10, March 1978.

LANCASTER, F. W. *Towards paperless information systems.* New York, Academic Press, 1978.

LE GOFF, J. *La civilisation de l'occident médiéval.* Paris, Arthaud, 1967.

LEBRETON, P. Les aspects écologiques des plantes nucleaires dans l'environnement côtier. *Bulletin d'Écologie*, 7(1): 33–59, 1976.

LEE, J. A. *Environmental security and global development: The essential connection.* Milwaukee, 1982.

LIMAYE, M. *Politics after freedom.* New Delhi, Atma Ram, 1982.

LINERR, M. Gift of poison: The unacceptable face of development aid. *Ambio*, 11(1): 2–8, 1982.

LINS DA SILVA, C. E. Jornalismo e ecologia. *Communicação e Sociedade*, 7: 54–55, Mar. 1982.

MACHADO, A. D. (ed.). *Energia nuclear e sociedade.* Rio de Janeiro, Paz e Terra, 1980.

MCLEOD, G. K. Some public health lessons from Three Mile Island: A case study in chaos. *Ambio*, 10(1): 18–23, 1981.

MADDOX, J. *The doomsday syndrome.* London, Macmillan, 1972.

MALTHUS, T. R. *Essay on the principle of population.* 7th ed. London, J. M. Dent, 1816.

MARCONI, P. *A censura política na imprensa brasileira (1968–1978).* São Paulo, Global, 1980.

MARTYN, J. *Report on the evaluation of INFOTERRA for the United Nations Environment Programme.* Paris, UNESCO, 1981.

MARX, K., and ENGELS, F. *Selected works.* London, Lawrence and Wishart, 1970.

MEADOWS, D. H., *et al. The limits to growth: A report for the Club of Rome's project on the predicament of mankind.* New York, Universe Books, 1972.

MESAROVIC, M., and PESTEL, E. *Mankind at the turning point: The second report to the Club of Rome.* New York, Dutton, 1974.

MEXICO. Comisión Intersecretarial de Saneamiento Ambiental. *México: Diez años después de Estocolmo.* Nairobi, 1982.

———. Subsecretaria de Mejoramiento del Ambiente. *Informe nacional gobierno de México.* Nairobi, PNUMA, 1982.

MICHELSON, W. *Environmental choice: Human behavior and residential satisfaction.* New York, Oxford University Press, 1977.

MUNN, R. F. Appropriate technology and information services in developing countries. *International Librarianship Review*, 10: 23–27, 1978.

MYERS, N. China's approach to environmental conservation. *Environmental Affairs*, 5(1): 33–63, Winter 1976.

NADER, R. *Ecotactics*. New York, Pocket Books, 1970.

NEELAMEGHAN, A., and SEETHARAMA, S. Information transfer: The next twenty-five years. *Library Science*, 13(1): 11–12, Mar. 1976.

NICHOLSON, J. M. Citizens can have a voice in environment decision-making. *Catalyst for Environment Energy*, 7(2): 28–31, 1980.

PADDOCK, W., and PADDOCK, P. *Famine—1975!* Boston, Little, Brown, 1967.

PALME, O. The cost of over-kill. *Uniterra*, 6(5): 13, Sept./Oct. 1981.

PARLOUR, J. W., and SCHATZOW, S. The mass media and public concern for environmental problems in Canada, 1960–1972. *International Journal of Environmental Studies*, 13(1): 9–17, 1978.

PEREIRA, L. *Ensaios de sociologia do desenvolvimento*. São Paulo, Pioneira, 1970.

———. *Urbanização e subdesenvolvimento*. 4th ed. Rio de Janeiro, Zahar, 1979.

POGELL, S. M. Government-initiated public participation in environmental decisions. *Environmental Comment*, 4: 4–6, Apr. 1979.

POPENOE, D. Urban sprawl: Some neglected sociologial considerations. *Sociology and Social Research*, 63(2): 255–68, 1979.

POQUET, G. The limits to global modelling. *International Social Science Journal*, 30(2): 284–300, 1978.

PREBISH, R. A crise do capitalismo maduro. *Novos Estudos Cebrap*, 2(1): 22, Apr. 1983.

QUIGLEY, C. Our ecological crisis. *Current History*, 59(347): 1–12, July 1970.

RAO, B. R. Concepto de ecología en la literatura védica. *Mazingira*, 6(4): 65–77, 1982.

RATTNER, H. Uma tecnologia para combater a pobreza. *Revista Brasileira de Tecnologia*, 12(2): 60–66, Apr./June 1981.

REYES MATTA, F. *A informação na nova ordem internacional*. Rio de Janeiro, Paz e Terra, 1980.

ROBERTSON, J. Towards post-industrial liberation and reconstruction. *New Universities Quarterly*, 32(1): 6–24, Winter 1977–78.

ROTHSCHILD (Lord). *Risk*. London, BBC, 1978 (Richard Dimbleby Lecture, Nov. 1978).

RUBIN, D. M. *et al*. Environmental information. *Annals of New York Academy of Sciences*, 216(7): 175–76, 1973.

RUGH, W. A. *The Arab press: News media and political process in the Arab World*. Syracuse, N.Y., Syracuse University Press, 1979.

160 Bibliography

SACHS, I. Environment and styles of development. *Environment in Africa*, 1(1): 9–33, Dec. 1974.

———. Sem medo de discordar. *Visão*, pp. 90–95, July 23, 1979.

SAGLIO, J. F. Implantation des centrales nucleaires et l'environment. *Annales des Mines*, pp. 139–44, Mar./Apr. 1976.

SANTOS, T. dos. *Imperialismo y dependencia*. Mexico, Era, 1978.

SAUVY, A. *Zero growth?* Oxford, England, B. Blackwell, 1975.

SCHMANDT, H. J., and BLOOMBERG, W., Jr. *The quality of urban life*. Beverly Hills, Calif., Sage, 1969.

SCHUMACHER, E. T. *Small is beautiful: A study of economics as if people mattered*. London, Abacus, 1973.

SIEHL, G. H. Our world—And welcome to it! *Library Journal*, 95: 1443–47, Apr. 15, 1970.

SILLS, D. L. (ed.). *Internatioal encyclopedia of the social sciences*. New York, Macmillan and Free Press, 1968.

SOMERVILLE, A. N. Academia and the environment: Environmental information. *Journal of Chemical Information and Computer Sciences*, 16(1): 1–4, 1976.

SORJ, B., *et al. Sociedade e política no Brasil pós–64*. São Paulo, Brasiliense, 1983.

STRETTON, H. *Capitalism, socialism and the environment*. Cambridge, Cambridge University Press, 1976.

SUNKEL, O. *La dimensión ambiental en los estilos de desarollo de América Latina*. Santiago de Chile, CEPAL, 1981.

TAAGEPERA, R. People, skills and resources: An interaction model for world population growth. *Technological Forecasting and Social Change*, 13(1): 13–30, Jan. 1979.

TALISAYON, S. D. New development goals and values in response to the global environmental crisis. *Science and Public Policy*, pp. 21–26, Feb. 1983.

TOLBA, M. The challenge of the eighties. *Uniterra*, 6(1): 11, Jan./Feb. 1981.

TRZYNA, T. C., and COAN, E. V. *World directory of environmental organizations*. 2d ed. Claremont, Calif., Public Affairs Clearinghouse, 1976.

UN. *Development and environment*. Report submitted by a panel of experts convened by the Secretary-General of the UN Conference on the Human Environment. Founex, 1971.

———. *International drinking water supply and sanitation decade: Report of the Secretary-General*. New York, 1980.

———. *Report of the conference on the human environment*. New York, 1973.

UNEP. *Annual report of the executive director, 1982*. Nairobi, 1983.

———. *Choosing the options: Alternative lifestyles and development patterns*. Nairobi, 1980.

———. *The Cocoyoc declaration*. Cocoyoc, Mexico, 1974.

————. *Directory of institutions and individuals active in environmentally sound and appropriate technologies.* Oxford, Pergamon Press, 1979.

————. Environment and development. *Facts*, p. 2, UNEP FS/13, Mar. 1974.

————. *The environment in 1982: Retrospect and prospect.* Nairobi, 1982 (UNEP/GC/SSC/2).

————. *Review of major achievements in the implementation of the Stockholm action plan.* Addendum: *Evaluation of the implementation of the 1982 goals.* Nairobi, 1981.

————. *The state of the environment, 1972–1982.* Nairobi, 1982.

————. *The state of the environment, 1980: Selected topics.* Nairobi, 1981.

————. *The state of the environment, 1981: Selected topics.* Nairobi, 1981.

UNESCO. *Un solo mundo, voces múltiples: Communicación e información en nuestro tiempo.* México, Fondo de Cultura, 1980.

UNIDO. *Long-term prospects of industrial development in the Arab Republic of Egypt.* Algiers, Fifth Industrial Development Conference for Arab States, 1980.

UNITED STATES. Council on Environmental Quality. *The global 2000 report to the president: Entering the twenty-first century.* Washington, D.C., U.S. Government Printing Office, 1980.

VITTACHI, T. The world population: Back from the brink? *The Guardian*, p. 17, Aug. 28, 1979.

WALSH, E. J. Three Mile Island: Meltdown of democracy? *Bulletin of the Atomic Scientists*, 39(3): 57–60, Mar. 1983.

WARD, B. *Progress for a small planet.* Middlesex, England, Penguin, 1979.

WEINBERG, A. M. Is nuclear power acceptable? *Science and Public Policy*, pp. 455–66, Oct. 1976.

WESTING, A. H. A world in balance. *Environmental Conservation*, 8(3): 183, Autumn 1981.

WHITE, L. Jr. The historical roots of our ecological crisis. *Science*, 155(3767): 1203, Mar. 10, 1967.

WHOLWILL, J. F. The social and political matrix of environmental attitudes and analysis of the vote on the California Coastal Zone Regulation Act. *Environment and Behavior*, 11(1): 71–85, Mar. 1979.

WILKINSON, R. G. *Poverty and progress: An ecological perspective on economic development.* New York, Praeger, 1973.

WYNNE, B. *Rationality and ritual: The Windscale inquiry and nuclear decisions in Britain.* Chalfont St. Giles, Bucks, British Society for the History of Science, 1982.

————. Redefining the issues of risk and public acceptance: The social viability of technology. *Futures*, pp. 13–31, Feb. 1983.

ZACHARIAS, J. R., *et al.* Common sense and nuclear peace. *Bulletin of the Atomic Scientists*, 39(4): Apr. 1983, special supplement.

Index

ABES, 90

Academy of Scientific Research and Technology, 96, 113

ACAST, 38

Afghanistan, 15

Africa: environmental education, 62; information systems, 95-9; soil degradation, 49-50; utilization of pesticides, 22

African Curriculum Organization, 62

African Regional Center for Technology, 95

African Rural Housing Association, 95

African Rural Storage Center, 96

Agricultural information systems, 87-88, 96, 97, 109, 112

AGRINTER, 87, 109, 112

Águas de Março, 61

Akioshi, T., 61

Al-Azhar University, 98

Allerton Park Institute, 132

Almeida, M. Ozório de, 25

Alternative life styles, 22, 122, 123

Alternative press, 130

Alternative technology, 13, 22, 25-26, 74, 95, 123; for information systems, 41-42, 91, 95

Althusser, L., 23

Amado, J., 61

Animism, 10

APPLE, 102

Appropriate technology: 13, 22, 25-26, 95, 123; for Health Information System, 41; for other information systems, 41, 91, 95

"Arab Science Abstracts", 98

Argentina, 62

Armaments race, 14-15, 38

ASFIS, 41

Asia: environmental education of, 62, 63-64; information systems of, 99-104; soil degradation of, 49-50

Asociación Interamericana de Ingeniería Sanitaria y Ambiental, 87

Associação Gaúcha de Proteção ao Ambiente Natural, 129

Atmospheric pollution, 37; information systems for, 91, 93, 94

Audubon Society, 57

Banaras Hindu University, 103

Bangladesh, 60

Bangladesh Agricultural Research Council, 99

About the Author

ANNA DA SOLEDADE VIEIRA is Professor at the Library School of the Universidade Federal de Minas Gerais in Brazil. Her earlier writings include *Environmental Information: An Approach to Pollution Control in Brazil,* as well as articles published in *Revista da Escola de Biblioteconomia da UFMG, Revista de Biblioteconomia de Brasília,* and *Ciência da Informação.*